HADAMARD MATRIX ANALYSIS AND SYNTHESIS

With Applications to Communications and Signal/Image Processing

THE KLUWER INTERNATIONAL SERIES IN ENGINEERING AND COMPUTER SCIENCE

HADAMARD MATRIX ANALYSIS AND SYNTHESIS

With Applications to Communications and Signal/Image Processing

by

R. K. Rao Yarlagadda
Oklahoma State University

John E. Hershey
General Electric Corporate Research & Development Center

KLUWER ACADEMIC PUBLISHERS
Boston / Dordrecht / London

Distributors for North America:
Kluwer Academic Publishers
101 Philip Drive
Assinippi Park
Norwell, Massachusetts 02061 USA

Distributors for all other countries:
Kluwer Academic Publishers Group
Distribution Centre
Post Office Box 322
3300 AH Dordrecht, THE NETHERLANDS

Library of Congress Cataloging-in-Publication Data

A C.I.P. Catalogue record for this book is available from the Library of Congress.

Printed on acid-free paper.

Printed in the United States of America

Dedication

To my Late parents,

Sri Yarlagadda China Ghantiah
Srimathi Yarlagadda Sobhanachalamma

to my brother-in-law and sister,

Dr. Gutta Kutumba Rao
Srimathi Gutta Narayanamma

and to my Late uncle and Late aunt,

Sri Korlipara Kasi Vishwanath Venkatachalam
Srimathi Korlipara Varalakshmamma

for their love, affection, and their understanding

— Rao Yarlagadda

To my wife and my best friend,

Anna

— John Hershey

Contents

PREFACE

There have been many excellent treatises on Hadamard matrices, transforms and applications such as that by Beauchamp (1984). These works encompass the spectrum of Hadamard and Walsh-Hadamard constructions. This tract, on the other hand, concerns only the Hadamard matrix which has become known as the *Sylvester* construction. It takes its name from J. J. Sylvester whose published work in 1867 was a harbinger of the Kronecker construction about which we write. For an excellent historical overview of Hadamard matrices and Sylvester's role (pre-Hadamard, of course), we recommend the survey paper by Craigen and Wallis (1993).

The applications we present that use the Sylvester construction touch on statistics, error correction coding theory, communications signaling, boolean function analysis and synthesis, image processing, sequence theory and signal representation. This is not to be considered an exhaustive list; quite to the contrary, in fact. It is a sampling of the exceptional importance of this simple Hadamard construction to a vast field of applications.

We start with an abbreviated review of Kronecker matrix algebra as it is the very basis of the Sylvester construction. It is also the key to the fast algorithm for implementation of the corresponding Hadamard transform. Next we define and explore some of the basic properties of the Sylvester constructed Hadamard matrix, particularly its eigenvectors and their construction. Following that we consider other constructions of the Sylvester form of the Hadamard matrix.

Next we define the Hadamard transform and elaborate on its fast implementation. We also look at relevant computational structures.

We then move on to applications where we first consider the use of the Hadamard matrix in statistics which was perhaps the first significant use of the fast Hadamard transform having been introduced for such purpose by the great statistician Yates in the mid nineteen thirties. We proceed then to a more nearly contemporary topic, that of error correction coding.

Next we look at some important issues involving boolean functions, their approximations and synthesis. This is followed by section on sequences which have proved to be useful to communications, radar and signal processing in general. These sequences of especial interest and with very close ties to Hadamard matrices are the maximal length binary sequences (*m-sequences*), composite sequences, and the Thue-Morse sequence.

Following this we take a look at a new signal representation and a new Hadamard basis and its use for communications signaling.

We conclude with an image encoding method termed the *Naturalness Preserving Transform* which provides a way to encode an image so that its reconstruction degrades gracefully as its transform is eroded by, perhaps, transmission through a very noisy transmission channel.

We apologize for, or at least admit in advance to, moving between the GF(2) additive and multiplicative groups, ie, operations using $\{0,1\}$ characters and $\{\pm 1\}$

characters. We believe it is unnatural to try to force one representation to all situations. Also, there are some instances where the field structure is necessary and thus involves both groups simultaneously.

We sincerely hope that the reader finds something that will be of help to a particular problem and also that the beauty of these gifts from the nineteenth century will inspire.

Rao Yarlagadda
John Hershey

Chapter 1 Kronecker Matrix Algebra

The study of the recursive Sylvester-constructed Hadamard matrix and transform is made considerably easier by employing Kronecker matrix algebra. In this short chapter we define the essential elements of this algebra and state, without proof, some useful theorems. There are a number of excellent references on Kronecker matrix algebra. The one we draw from is Regalia and Mitra (1989).

Let matrix A be of dimension $m \times n$ and matrix B of dimension $k \times l$. The Kronecker product of matrices A and B is a matrix of dimension $mk \times nl$ and is denoted as $A \otimes B$. It is defined as

$$A \otimes B = \begin{pmatrix} Ab_{0,0} & Ab_{0,1} & \cdots & Ab_{0,l-1} \\ Ab_{1,0} & Ab_{1,1} & \cdots & Ab_{1,l-1} \\ \vdots & \vdots & \vdots & \vdots \\ Ab_{k-1,0} & Ab_{k-1,1} & \cdots & Ab_{k-1,l-1} \end{pmatrix} \tag{1}$$

where $b_{i,j}$ denotes the i,j-th element of matrix B.

The following are key properties of the Kronecker matrix product:

i.

$$(A \otimes B) \otimes C = A \otimes (B \otimes C) \tag{2}$$

ii.

$$(A \otimes B)(C \otimes D) = (AC) \otimes (BD) \tag{3}$$

where C is of dimension $n \times q$ and D is of dimension $l \times s$

iii.

$$\begin{aligned} A \otimes D &= (AI_n) \otimes (I_l D) \\ &= (A \otimes I_l)(I_n \otimes D) \end{aligned} \tag{4}$$

where I_j denotes a $j \times j$ identity matrix

iv.

$$(A \otimes B)^T = A^T \otimes B^T \tag{5}$$

v.

$$|A \otimes B| = |A|^r |A|^s \tag{6}$$

where A and B are square matrices of dimensions $r \times r$ and $s \times s$ respectively.

We shall also make use of the *direct sum* of matrices. This operation is denoted by $\oplus$. Again, following Regalia and Mitra (1989), let there be N matrices, $A_0, A_1, \ldots, A_{N-1}$. These matrices *need not* be of the same dimensions. We define their direct sum as

$$A_0 \oplus A_1 \oplus \cdots \oplus A_{N-1} = \begin{pmatrix} A_0 & & & \\ & A_1 & & \\ & & \ddots & \\ & & & A_{N-1} \end{pmatrix}. \tag{7}$$

The direct sum has the following properties

i.

$$(A_0 A_1) \oplus (A_2 A_3) = (A_0 \oplus A_2)(A_1 \oplus A_3) \tag{8}$$

ii.

$$|A_0 \oplus A_1 \oplus \cdots A_{N-1}| = \prod_{i=0}^{N-1} |A_i| \tag{9}$$

if each A_i is a square matrix.

iii. The direct sum, $A_0 \oplus A_1$ of two orthogonal matrices, A_0 and A_1, is orthogonal.

Chapter 2 A Hadamard Matrix

A Hadamard matrix, $H = (h_{ij})$ is defined as a square matrix of dimension nxn where:

i. All entries are ± 1,
ii. Any two distinct rows are orthogonal, ie, $\forall\ i, j,\ i \neq j$, $\sum_k h_{ik} h_{jk} = 0$.

The above two conditions require that n be at least even for a Hadamard matrix to exist.

Arguing further, as per Harwit and Sloane (1979), we consider a Hadamard matrix for which $n > 2$. Any row of such a matrix, excepting the all ones row (or all minus ones row), must have exactly n/2 ones and n/2 zeros. For orthogonality to obtain, for any two distinct rows, i and j, not involving the all ones (or all minus ones) row, there must exist n/4 columns where the elements of rows i and j are both +1, n/4 columns where the elements of rows i and j are +1 and -1 respectively, n/4 columns where the elements of rows i and j are -1 and +1 respectively, and n/4 columns where the elements of rows i and j are both -1. Thus, a Hadamard matrix cannot exist for an n greater than 2 which is not a multiple of 4.

Chapter 3 The Sylvester-Hadamard Matrix of Rank 2^n

We will be concerned with a particular form of the Hadamard matrix of rank 2^n. This form is produced using a recursive Kronecker product. Specifically, the Hadamard matrix of interest is designated a *Sylvester-Hadamard* Matrix after Sylvester (1867), denoted as H_n, and created by

$$H_n = H_1 \otimes H_{n-1} \tag{10}$$

where

$$H_1 = \begin{pmatrix} +1 & +1 \\ +1 & -1 \end{pmatrix}. \tag{11}$$

Thus,

$$H_2 = \begin{pmatrix} +1 & +1 & +1 & +1 \\ +1 & -1 & +1 & -1 \\ +1 & +1 & -1 & -1 \\ +1 & -1 & -1 & +1 \end{pmatrix} \tag{12}$$

and

$$H_3 = \begin{pmatrix} +1 & +1 & +1 & +1 & +1 & +1 & +1 & +1 \\ +1 & -1 & +1 & -1 & +1 & -1 & +1 & -1 \\ +1 & +1 & -1 & -1 & +1 & +1 & -1 & -1 \\ +1 & -1 & -1 & +1 & +1 & -1 & -1 & +1 \\ +1 & +1 & +1 & +1 & -1 & -1 & -1 & -1 \\ +1 & -1 & +1 & -1 & -1 & +1 & -1 & +1 \\ +1 & +1 & -1 & -1 & -1 & -1 & +1 & +1 \\ +1 & -1 & -1 & +1 & -1 & -1 & +1 & -1 \end{pmatrix}. \tag{13}$$

Chapter 4 The Eigenvalues of H_n

The order of H_n is 2^n and the 2^n eigenvalues of H_n are distributed as:

i. 2^{n-1} eigenvalues are $+2^{n/2}$,
ii. 2^{n-1} eigenvalues are $-2^{n/2}$.

The proof of this assertion is straightforward. Consider that $\frac{1}{\sqrt{N}}H_n$ is a symmetric orthogonal matrix and therefore the eigenvalues of $\frac{1}{\sqrt{N}}H_n$ are ± 1 and there are 2^n of them. The trace of H_n is zero and thus there must be exactly 2^{n-1} ones and 2^{n-1} minus ones.

Chapter 5 The Eigenvectors of H_n

The eigenvectors of H_n can be constructed in many ways. One method is to use a modal matrix. Following Yarlagadda and Hershey (1982a) we first define a modal matrix, T_n, of H_n by

$$T_n = H_n + \Lambda_n \tag{14}$$

where Λ_n is a diagonal matrix of the eigenvalues of H_n as suggested by Kremer (1973) who continued on to suggest Gram-Schmidt orthogonalization. Yarlagadda and Hershey (1982a) felt that a more efficient method might obtain if the problem could be set as an eigenvector problem of symmetric idempotent matrices as described in Householder and Fox (1971). The motivation is in noting that because

$$F_1 = \frac{1}{2}\left(I + 2^{-n/2} H_n\right) \tag{15}$$

and

$$F_2 = -\frac{1}{2}\left(I - 2^{-n/2} H_n\right) \tag{16}$$

are idempotent matrices, the eigenvectors of $2^{-n/2} H_n$ can be found through decomposition, eg, Cholesky's representation:

$$F_1 = G_1 G_1^T\,, \tag{17}$$

$$F_2 = G_2 G_2^T\,. \tag{18}$$

A modal matrix of $2^{-n/2} H_n$ is then

$$T_n = (G_1\; G_2)\,, \tag{19}$$

and, as Householder and Fox (1971) note, the columns of T_n are orthogonal.

Now let X_n be an eigenvector of H_n corresponding to the positive eigenvalue $2^{n/2}$. Then

$$X_{n+1} = \begin{pmatrix} X_n \\ (-1+\sqrt{2})X_n \end{pmatrix} \tag{20}$$

is an eigenvector of H_{n+1} corresponding to the positive eigenvalue $2^{(n+1)/2}$. This can be easily seen by observing that

$$\begin{aligned} \begin{pmatrix} H_n & H_n \\ H_n & -H_n \end{pmatrix} \begin{pmatrix} X_n \\ (-1+\sqrt{2})(X_n) \end{pmatrix} &= \begin{pmatrix} \sqrt{2} H_n X_n \\ (2-\sqrt{2}) H_n X_n \end{pmatrix} \\ &= 2^{(n+1)/2} \begin{pmatrix} X_n \\ (-1+\sqrt{2})(X_n) \end{pmatrix} \end{aligned} \tag{21}$$

X_1 can be easily found to be

$$X_1 = \begin{pmatrix} 1 \\ -1+\sqrt{2} \end{pmatrix} . \tag{22}$$

From (20) it is clear that X_n can be written as

$$X_n = (a_0 + \sqrt{2}b_0 \quad a_1 + \sqrt{2}b_1 \quad a_2 + \sqrt{2}b_2 \quad \dots \quad a_{2^n-1} + \sqrt{2}b_{2^n-1})^T , \tag{23}$$

where the $\{a_i, b_i\}$ are integer pairs. X_n has the general form

$$X_n = (q^0 \quad q^1 \quad q^1 \quad q^2 \quad q^1 \quad \dots \quad q^n)^T , \tag{24}$$

where $q = -1 + \sqrt{2}$.

A very important property has emerged and is that

$$a_{2k+1} + \sqrt{2}b_{2k+1} = \left(-1+\sqrt{2}\right)\left(a_{2k} + \sqrt{2}b_{2k}\right) . \tag{25}$$

By equating rational and irrational components, we see that

$$\begin{aligned} a_{2k+1} &= 2b_{2k} - a_{2k} , \\ b_{2k+1} &= a_{2k} - b_{2k} . \end{aligned} \tag{26}$$

One further gift from (20) is that

$$a_{(2^i)k} + \sqrt{2}b_{(2^i)k} = a_k + \sqrt{2}b_k \tag{27}$$

for all integers $i \geq 0$. We may now compute the components of X_n without expanding the powers of q.

We now define a sequence of matrices of order $2^n \times 2^n$, E_n, recursively as follows:

$$\begin{aligned} E_0 &= (-1) \\ E_{n+1} &= \begin{pmatrix} 0 & E_n \\ -E_n & 0 \end{pmatrix} , \; n \geq 0, \end{aligned} \tag{28}$$

where 0 stands for the zero matrix of order $2^n \times 2^n$. We note that E_n is symmetric for n even and skew-symmetric for n odd.

We can now state that

$$E_n H_n = (-1)^n H_n E_n . \tag{29}$$

To prove this assertion, note that by induction

$$E_n^2 = (-1)^n I_{2^n} \tag{30}$$

and that

$$
\begin{aligned}
E_0 H_0 E_0 &= H_0 \, , \\
E_1 H_1 E_1 &= H_1 \, .
\end{aligned}
\tag{31}
$$

Further,

$$
\begin{aligned}
E_{n+1} H_{n+1} E_{n+1} &= \begin{pmatrix} 0 & E_n \\ -E_n & 0 \end{pmatrix} \begin{pmatrix} H_n & H_n \\ H_n & -H_n \end{pmatrix} \begin{pmatrix} 0 & E_n \\ -E_n & 0 \end{pmatrix} \\
&= \begin{pmatrix} E_n H_n E_n & E_n H_n E_n \\ E_n H_n E_n & -E_n H_n E_n \end{pmatrix} \\
&= H_{n+1} \, ,
\end{aligned}
\tag{32}
$$

thus proving the assertion.

Next, we define for each $k = 0, 1, ..., n$, a matrix P_{nk} as the direct sum of 2^{n-k} copies of E_k, ie, $P_{nk} = E_k \oplus E_k \oplus \cdots \oplus E_k$. Note that

$$
P_{nk} = \begin{cases} P_{nk}^T & , k \ even, \\ -P_{nk}^T & , k \ odd. \end{cases}
\tag{33}
$$

We can now state that

$$
P_{nk} H_n P_{nk}^T = (-1)^k H_n \, .
\tag{34}
$$

To see this, perform the block multiplication

$$
\begin{pmatrix} E_k & & & \\ & E_k & & \\ & & \ddots & \\ & & & E_k \end{pmatrix} \begin{pmatrix} H_k & H_k & \cdots & H_k \\ H_k & -H_k & \cdots & -H_k \\ \vdots & \vdots & & \vdots \\ H_k & -H_k & \cdots & \pm H_k \end{pmatrix}
\cdot \begin{pmatrix} E_k^T & & & \\ & E_k^T & & \\ & & \ddots & \\ & & & E_k^T \end{pmatrix} .
$$

Each block is $E_k(\pm H_k)E_k^T = (-1)^k(\pm H_k)$ and the result follows.

One further bit of foundation is to note the commutation

$$
P_{nk} P_{nj} = P_{nj} P_{nk} \, , \ \forall k.
\tag{35}
$$

The proof is straightforward. We let $n > k > j$ and form

$$
\begin{aligned}
P_{nk} P_{nj} &= \begin{pmatrix} E_k & & \\ & \ddots & \\ & & E_k \end{pmatrix} \begin{pmatrix} E_j & & \\ & \ddots & \\ & & E_j \end{pmatrix} \\
&= \begin{pmatrix} E_k(E_j \oplus \cdots \oplus E_j) & & \\ & \ddots & \\ & & E_k(E_j \oplus \cdots \oplus E_j) \end{pmatrix} .
\end{aligned}
\tag{36}
$$

All we now need to show is the commutation $E_k(E_j \oplus \cdots \oplus E_j) = (E_j \oplus \cdots \oplus E_j)E_k$. This can be seen by using a previous recursion, ie,E_k can be expressed in terms of E_j (j<n) as

$$E_k = \begin{pmatrix} & & & \pm E_j \\ & & \pm E_j & \\ & \cdot^{\cdot^{\cdot}} & & \\ \pm E_j & & & \end{pmatrix} \tag{37}$$

where the explicit sign of E_j can be determined from the recursion. Therefore

$$E_k(E_j \oplus \cdots \oplus E_j) = \begin{pmatrix} & & E_j^2 \\ & \ddots & \\ E_j^2 & & \end{pmatrix} \tag{38}$$

and the result follows. (The result also follows for n=k per (35).)

We are now ready to generate a set of orthogonal eigenvectors. Recall that we have already computed an eigenvector for H_n, ie, X_n in (24). For notational simplicity, let $Y_1 = X_n$. Consider the set of vectors $\{Y_i\}$, $2 \leq i \leq 2^n$, formed by the recursion

$$Y_i = Q_{i-1}Y_{i-1} \ , \ 2 \leq i \leq 2^n \ , \tag{39}$$

where

$$Q_{2^r(2k+1)} = P_{n(r+1)} \ , \ 0 \leq k \, . \tag{40}$$

The $\{Y_i\}$ are eigenvectors. This is seen by noting that $Y_i = P_\pi Y_1$ where P_π is the appropriate product of the P_{nk} matrices. We now premultiply Y_i by H_n and use (34).

$$\begin{aligned} H_n Y_i &= H_n P_\pi Y_1 \\ &= \lambda P_\pi H_n Y_1 \\ &= \lambda P_\pi Y_1 \\ &= \lambda Y_i \end{aligned} \tag{41}$$

where λ takes care of the signs in using (34). (Note that the matrix$\frac{1}{||Y_1||}(Y_1 Y_2 \cdots Y_{2^n})$ is an orthogonal modal matrix of H_n where $||Y_1|| = \left(4 - \sqrt{2}2\right)^{n/2}$ for $n \geq 2$.)

It is now clear that each entry of Y_1 will be shifted out of its original location and successively shifted into and out of the remaining $2^n - 1$ locations by successive applications of the recursion given by (39). This can be intuitively grasped by considering the motion of any selected row element of Y_1. For example, the zeroth element, unity, undergoes the circuit of position numbers $0 \to 1 \to 2 \to 3 \to \cdots \to$

$2^n - 1$. For an illustration of the circuit behavior, consider the progression of vectors $Y_1, Y_2, ..., Y_8$:

$$
\begin{aligned}
Y_1^T &= \left(1, -1+\sqrt{2}, -1+\sqrt{2}, 3-2\sqrt{2}, -1+\sqrt{2}, 3-2\sqrt{2}, 3-2\sqrt{2}, -7+5\sqrt{2}\right) \\
Y_2^T &= \left(-\sqrt{2}+1, 1, 2\sqrt{2}-3, -1+\sqrt{2}, 2\sqrt{2}-3, -1+\sqrt{2}, 7-5\sqrt{2}, 3-2\sqrt{2}\right) \\
Y_3^T &= \left(-\sqrt{2}+1, 2\sqrt{2}-3, 1, -1+\sqrt{2}; 2\sqrt{2}-3, 7-5\sqrt{2}, -1+\sqrt{2}, 3-2\sqrt{2}\right) \\
Y_4^T &= \left(3-2\sqrt{2}, -\sqrt{2}+1, 1-\sqrt{2}, 1, -7+5\sqrt{2}, 2\sqrt{2}-3, 2\sqrt{2}-3, -1+\sqrt{2}\right) \\
Y_5^T &= \left(-\sqrt{2}+1, 2\sqrt{2}-3, 2\sqrt{2}-3, 7-5\sqrt{2}, 1, \sqrt{2}-1, \sqrt{2}-1, 3-2\sqrt{2}\right) \\
Y_6^T &= \left(3-2\sqrt{2}, 1-\sqrt{2}, -7+5\sqrt{2}, 2\sqrt{2}-3, 1-\sqrt{2}, 1, 2\sqrt{2}-3, \sqrt{2}-1\right) \\
Y_7^T &= \left(3-2\sqrt{2}, -7+5\sqrt{2}, 1-\sqrt{2}, 2\sqrt{2}-3, 1-\sqrt{2}, 2\sqrt{2}-3, 1, \sqrt{2}-1\right) \\
Y_8^T &= \left(7-5\sqrt{2}, 3-2\sqrt{2}, 3-2\sqrt{2}, 1-\sqrt{2}, 3-2\sqrt{2}, 1-\sqrt{2}, 1-\sqrt{2}, 1\right)
\end{aligned}
\tag{42}
$$

We can now establish that

$$Y_i^T Y_j = 0 \ , \ i \neq j. \tag{43}$$

To prove this, first note that $Y_i^T Y_j$ may be written in the form

$$Y_i^T Y_j = Y_1^T (P_{ni_1} P_{ni_2} ... P_{ni_m}) Y_1 \ . \tag{44}$$

To show this intermediate step, we let $k = max(i_1, i_2, ..., i_m)$ and then recall the commutation property (35) and pick $k = i_1$[1] allowing us to write

$$P_{ni_1} P_{ni_2} ... P_{ni_m} = \begin{pmatrix} E_k B_k & & & \\ & E_k B_k & & \\ & & \ddots & \\ & & & E_k B_k \end{pmatrix} \tag{45}$$

where

$$B_k = \begin{pmatrix} E_{i_2} & & & \\ & E_{i_2} & & \\ & & \ddots & \\ & & & E_{i_2} \end{pmatrix} \cdots \begin{pmatrix} E_{i_m} & & & \\ & E_{i_m} & & \\ & & \ddots & \\ & & & E_{i_m} \end{pmatrix} . \tag{46}$$

Now Y_1 can be expressed as

$$Y_1 = \begin{pmatrix} D_k & & & \\ & D_k & & \\ & & \ddots & \\ & & & D_k \end{pmatrix} \begin{pmatrix} Z_{k1} \\ Z_{k2} \\ \vdots \\ Z_{km} \end{pmatrix} \tag{47}$$

[1] Note that $P_{ni_1} P_{ni_2} ... P_{ni_m}$ can be represented by a product of matrices: $-I, P_{n1} P_{n2} ... P_{nn}$. The first, $-I$, is introduced because it obtains upon squaring and subsequent reduction of terms like P_{n1}.

where $m = 2^{n-k}$ and the vector Z_{kj} is of order 2^k. The matrix D_k is defined recursively by

$$D_k = \begin{pmatrix} D_{k-1} & 0 \\ 0 & D_{k-1} \end{pmatrix}, \; 2 \le k \,, \tag{48}$$

with

$$D_1 = \begin{pmatrix} 1 & 0 \\ 0 & -1+\sqrt{2} \end{pmatrix}. \tag{49}$$

Given the structure of Y_1, ie, $Y_1 = X_n$ and the structure of D_k, it can be seen that the vector Z_{kj} may be split into two parts

$$Z_{kj} = \begin{pmatrix} Z'_{kj} \\ (-1+\sqrt{2})\, Z'_{kj} \end{pmatrix}. \tag{50}$$

We can now write (44) as

$$Y_1^T (P_{ni_1} \ldots P_{ni_m}) Y_1 = \sum_{j=1}^{m} Z_{kj}^T D_k E_k B_k D_k Z_{kj} \,. \tag{51}$$

Further, $E_k B_k$ can be written as

$$\begin{aligned} E_k B_k &= \begin{pmatrix} 0 & E_{k-1} \\ -E_{k-1} & 0 \end{pmatrix} \begin{pmatrix} B_{k-1} & 0 \\ 0 & B_{k-1} \end{pmatrix} \\ &= \begin{pmatrix} 0 & E_{k-1}B_{k-1} \\ -E_{k-1}B_{k-1} & 0 \end{pmatrix}. \end{aligned} \tag{52}$$

Now we may write

$$\begin{aligned} Z_{kj}^T D_k E_k B_k D_k Z_{kj} &= (\, Z_{(k-1)j}^T \quad (-1+\sqrt{2}) Z_{(k-1)j}^T \,) \\ &\cdot \begin{pmatrix} 0 & D_{k-1}E_{k-1}B_{k-1}D_{k-1} \\ -D_{k-1}E_{k-1}B_{k-1}D_{k-1} & 0 \end{pmatrix} \\ &\cdot \begin{pmatrix} Z_{(k-1)j} \\ (-1+\sqrt{2}) Z_{(k-1)j} \end{pmatrix} \\ &= \left(-1+\sqrt{2}\right) Z_{(k-1)j}^T D_{k-1}E_{k-1}B_{k-1}D_{k-1} Z_{(k-1)j} \\ &\quad - \left(-1+\sqrt{2}\right) Z_{(k-1)j}^T D_{k-1}E_{k-1}B_{k-1}D_{k-1} Z_{(k-1)j} \,. \end{aligned} \tag{53}$$

From the properties of P_{nk} it follows that

$$\left(E_{k-1}B_{k-1}\right)^T = \mp E_{k-1}B_{k-1} \,. \tag{54}$$

In the first case, each term in (51) is zero (as $\alpha^T A \alpha = 0$ when A is skew-symmetric). In the second case, both terms in (53) are equal and therefore (51) also exhibits a null sum. Therefore, in all cases, the summation in (51) is zero and therefore (51) is zero which establishes the orthogonality.

Chapter 6 Other Constructions of H_n

Ahmed and Rao (1975) provide an exponential definition for H_n by writing it directly as

$$\begin{aligned} H_n &= (h_{ij}) \\ &= (-1)^{\langle i,j \rangle} \end{aligned} \tag{55}$$

where $\langle i, j \rangle$ is the modulo-two inner product of the two normal binary representations of i and j. Table 1 shows calculation of the inner product for H_2.

Table 1 Direct Construction of H_2; inner products (left), $(-1)^{\langle i,j \rangle}$ (right)

		j					h_{ij}			
		0,0	0,1	1,0	1,1					
	0,0	0	0	0	0		+1	+1	+1	+1
	0,1	0	1	0	1		+1	-1	+1	-1
i	1,0	0	0	1	1	$\leftrightarrow$	+1	+1	-1	-1
	1,1	0	1	1	0		+1	-1	-1	+1

Henderson (1970) recorded a nice factorization of H_n. We construct B_n, of order $2^n \times n$, and is the matrix representing an n-bit counter counting, using zeros and ones, top to bottom, in normal binary starting at zero. Thus

$$B_1 = \begin{pmatrix} 0 \\ 1 \end{pmatrix} \tag{56}$$

and

$$B_2 = \begin{pmatrix} 0 & 0 \\ 0 & 1 \\ 1 & 0 \\ 1 & 1 \end{pmatrix} . \tag{57}$$

To obtain H_n, we form $B_n B_n^T$ modulo-two and then change the ones to minus ones and the zeros to ones, eg,

$$\begin{aligned} B_1 B_1^T &= \begin{pmatrix} 0 \\ 1 \end{pmatrix} \begin{pmatrix} 0 & 1 \end{pmatrix} \\ &= \begin{pmatrix} 0 & 0 \\ 0 & 1 \end{pmatrix} \leftrightarrow \begin{pmatrix} +1 & +1 \\ +1 & -1 \end{pmatrix} \end{aligned} \tag{58}$$

and

$$
\begin{aligned}
B_2 B_2^T &= \begin{pmatrix} 0 & 0 \\ 0 & 1 \\ 1 & 0 \\ 1 & 1 \end{pmatrix} \begin{pmatrix} 0 & 0 & 1 & 1 \\ 0 & 1 & 0 & 1 \end{pmatrix} \\
&= \begin{pmatrix} 0 & 0 & 0 & 0 \\ 0 & 1 & 0 & 1 \\ 0 & 0 & 1 & 1 \\ 0 & 1 & 1 & 0 \end{pmatrix} \leftrightarrow \begin{pmatrix} +1 & +1 & +1 & +1 \\ +1 & -1 & +1 & -1 \\ +1 & +1 & -1 & -1 \\ +1 & -1 & -1 & +1 \end{pmatrix} .
\end{aligned} \tag{59}
$$

Another recursive factorization into sparse matrices is due to Lee and Kaveh (1986). Their recursion rests on a property of Kronecker matrix algebra which is given in Chapter 1. Their recursion rests on the identity

$$H_n = (H_1 \otimes I_{2^{n-1}})(I_2 \otimes H_{n-1}) \tag{60}$$

where I_k is an identity matrix of order k.[2]
By using (3) and (10) it follows that

$$(H_1 \otimes I_{2^{n-1}})(I_2 \otimes H_{n-1}) = H_1 \otimes H_{n-1} . \tag{61}$$

For purposes of developing a general form, we write (60) as

$$H_n = (I_1 \otimes H_1 \otimes I_{2^{n-1}})(I_2 \otimes H_{n-1}) \tag{62}$$

where I_1 is scalar unity.

The second term of (62), $I_2 \otimes H_{n-1}$ can be written as

$$
\begin{aligned}
I_2 \otimes H_{n-1} &= \begin{pmatrix} I_1 \otimes H_1 \otimes I_{2^{n-2}} & 0 \\ 0 & I_1 \otimes H_1 \otimes I_{2^{n-2}} \end{pmatrix} \begin{pmatrix} I_2 \otimes H_{n-2} & 0 \\ 0 & I_2 \otimes H_{n-2} \end{pmatrix} \\
&= (I_2 \otimes H_1 \otimes I_{2^{n-2}})(I_4 \otimes H_{n-2})
\end{aligned} \tag{63}
$$

and the decomposition becomes clear. We have

$$H_n = \prod_{i=1}^{n} (I_{2^{i-1}} \otimes H_1 \otimes I_{2^{n-i}}) . \tag{64}$$

As H_n is symmetric, it follows that $H_n^T = H_n$ and we also have that

$$H_n = \prod_{i=1}^{n} (I_{2^{n-i}} \otimes H_1 \otimes I_{2^{i-1}}) . \tag{65}$$

[2] Remember our convention that the subscript of the Hadamard matrix conveys the *log* (base two) of the order.

Chapter 7 The Hadamard Transform and Its Fast Implementation

The Hadamard transform, y, of a $2^n \times 1$ vector x is defined as

$$y = H_n x \,. \tag{66}$$

Straightforward calculation of (66) requires $O(2^{2n})$ arithmetic operations. There are faster methods just as there are faster methods for calculating similar transforms such as the Fourier and Discrete cosine transforms.

Most of these fast algorithms appeal to the decomposition

$$H_k = \begin{pmatrix} I_{2^{k-1}} & I_{2^{k-1}} \\ I_{2^{k-1}} & -I_{2^{k-1}} \end{pmatrix} \begin{pmatrix} H_{k-1} & 0_{2^{k-1}} \\ 0_{2^{k-1}} & H_{k-1} \end{pmatrix} \tag{67}$$

where $I_{2^{k-1}}$ is an identity matrix and $0_{2^{k-1}}$ the zero matrix, both of order 2^{k-1}.

Following Yarlagadda and Hershey (1981), let us now define

$$C_{n-k} = \begin{pmatrix} I_{2^{n-k}} & I_{2^{n-k}} \\ I_{2^{n-k}} & -I_{2^{n-k}} \end{pmatrix} , \; 0 \le k \le n \tag{68}$$

and

$$\begin{aligned} B_n &= C_{n-1} \\ B_{n-1} &= C_{n-2} \oplus C_{n-2} \\ &\vdots \\ B_1 &= C_0 \oplus C_0 \oplus \cdots C_0 \end{aligned} \tag{69}$$

where $\oplus$ is the direct sum. Note that H_n can be written as

$$H_n = B_n B_{n-1} \cdots B_1 \,. \tag{70}$$

The matrix H_n and the matrices $\{B_j\}$, $j = 1, 2, ..., n-1$ are symmetric matrices. Therefore

$$\begin{aligned} H_n &= B_n B_{n-1} \cdots B_1 \\ &= B_1 B_2 \cdots B_{n-1} \\ &= H_n^T \end{aligned} \tag{71}$$

resulting in another decomposition of H_n. Shortly, we will consider another decomposition that can be expressed H_n as a power of a matrix. For the nonce, note also that

$$H_n^{-1} = \frac{1}{N} H_n \,, \tag{72}$$

that is, $\frac{1}{\sqrt{N}}H_n$ is an orthogonal matrix. Also, $\frac{1}{\sqrt{2}}B_j$ is an orthogonal matrix.

We are now in a position to produce a very useful n-th root of H_n. We have

$$H_n = A^n \tag{73}$$

where

$$A = \begin{pmatrix} 1 & 1 & 0 & 0 & \cdots & 0 & 0 \\ 0 & 0 & 1 & 1 & \cdots & 0 & 0 \\ & & & & \vdots & & \\ 0 & 0 & 0 & 0 & \cdots & 1 & 1 \\ 1 & -1 & 0 & 0 & \cdots & 0 & 0 \\ 0 & 0 & 1 & -1 & \cdots & 0 & 0 \\ & & & & \vdots & & \\ 0 & 0 & 0 & 0 & \cdots & 1 & -1 \end{pmatrix} \tag{74}$$

The proof of this assertion is built on three well known properties of permutation matrices:

i. post multiplication of a matrix by a permutation matrix simply permutes the columns of that matrix,
ii. premultiplication of a matrix by a permutation matrix simply permutes the rows,
iii. the inverse of a permutation matrix is its transpose.

We write

$$H_n = B_n P_{n-1}\left(P_{n-1}^T B_{n-1} P_{n-2}\right)\left(P_{n-2}^T B_{n-2} P_{n-3}\right)\cdots\left(P_1^T B_1\right) \tag{75}$$

where the $\{P_i\}$ are permutation matrices. Next we will see that

$$\begin{aligned} A &= P_1^T B_1 \\ A &= P_{n-k}^T B_{n-k} P_{n-k-1} \ , \ k = 1, ..., n-2 \\ A &= B_n P_{n-1} \end{aligned} \tag{76}$$

with

$$\begin{aligned} P_{n-k}^T &= \left(P_1^T\right)^{n-k} \\ &= (P)^{n-k} \end{aligned} \tag{77}$$

and

$$P^n = I \ , \tag{78}$$

an identity matrix.

Let us define the permutation matrix, P, in a manner similar to that in Bellanger (1984). It is a transformation of an $N = 2^n$ dimensional vector

$$x^T = \left(x_0 x_1 x_2 \cdots x_{\frac{N}{2}-1} \vdots x_{\frac{N}{2}} x_{\frac{N}{2}+1} \cdots x_{N-2} x_{N-1}\right) \tag{79}$$

such that

$$Px = y \tag{80}$$

where

$$y^T = \left(x_0 x_{\frac{N}{2}} x_1 x_{\frac{N}{2}+1} \cdots x_{\frac{N}{4}+\frac{N}{2}-1} \vdots x_{\frac{N}{4}} x_{\frac{N}{4}+\frac{N}{2}} \cdots x_{\frac{N}{2}-1} x_{N-1} \right). \tag{81}$$

The $N \times N$ permutation matrix P can be explicitly written as a concatenation of two $N \times \frac{N}{2}$ matrices as

$$P = \begin{pmatrix} 1 & 0 & \cdots & 0 & 0 & 0 & \cdots & 0 & 0 \\ 0 & 0 & \cdots & 0 & 1 & 0 & \cdots & 0 & 0 \\ 0 & 1 & \cdots & 0 & 0 & 0 & \cdots & 0 & 0 \\ 0 & 0 & \cdots & 0 & 0 & 1 & \cdots & 0 & 0 \\ \vdots & \vdots & \vdots & \vdots & \vdots & \vdots & \vdots & \vdots & \vdots \\ 0 & 0 & \cdots & 0 & 0 & 0 & \cdots & 1 & 0 \\ 0 & 0 & \cdots & 1 & 0 & 0 & \cdots & 0 & 0 \\ 0 & 0 & \cdots & 0 & 0 & 0 & \cdots & 0 & 1 \end{pmatrix} \tag{82}$$

It is easy to see that $P^n = I$, an identity matrix, as n rotations result in the original vector. Therefore, $PP^{n-1} = I$ or $P^{-1} = P^T = P^{n-1}$.

As an example, we have, for $N = 16$,

$y^T = \left(x_0 x_8 x_1 x_9 x_2 x_{10} x_3 x_{11} \vdots x_4 x_{12} x_5 x_{13} x_6 x_{14} x_7 x_{15} \right)$ and

$x^T = \left(x_0 x_1 x_2 x_3 x_4 x_5 x_6 x_7 \vdots x_8 x_9 x_{10} x_{11} x_{12} x_{13} x_{14} x_{15} \right)$.

Next, it is true that $P^T B_1 = A$ where A is given in (74). This can be seen from

$$\begin{pmatrix} 1 & 0 & 0 & 0 & \cdots & 0 & 0 \\ 0 & 0 & 1 & 0 & \cdots & 0 & 0 \\ \vdots & \vdots & \vdots & \vdots & \vdots & \vdots & \vdots \\ 0 & 0 & 0 & 0 & \cdots & 1 & 0 \\ 0 & 1 & 0 & 0 & \cdots & 0 & 0 \\ 0 & 0 & 0 & 1 & \cdots & 0 & 0 \\ \vdots & \vdots & \vdots & \vdots & \vdots & \vdots & \vdots \\ 0 & 0 & 0 & 0 & \cdots & 0 & 1 \end{pmatrix} \begin{pmatrix} 1 & 1 & & 0 & 0 \\ 1 & -1 & & 0 & 0 \\ \vdots & \vdots & \vdots & \vdots & \vdots \\ 0 & 0 & & 1 & 1 \\ 0 & 0 & & 1 & -1 \end{pmatrix}$$

$$= \begin{pmatrix} 1 & 1 & 0 & 0 & \cdots & 0 & 0 \\ 0 & 0 & 1 & 1 & \cdots & 0 & 0 \\ \vdots & \vdots & \vdots & \vdots & \vdots & \vdots & \vdots \\ \cdots & \cdots & \cdots & \cdots & \cdots & 1 & 1 \\ -1 & -1 & 0 & 0 & \cdots & 0 & 0 \\ 0 & 0 & -1 & -1 & \cdots & & \\ \vdots & \vdots & \vdots & \vdots & \vdots & \vdots & \vdots \\ 0 & 0 & 0 & 0 & \cdots & 1 & -1 \end{pmatrix} \tag{83}$$

Premultiplication of B_1 by P^T permutes alternate rows of B_1 to the top and bottom half of the matrix A respectively. It is interesting to note that $P^T B_2 P = B_1$ which can be seen from

$$
\begin{pmatrix}
1 & 0 & 0 & 0 & \cdots & 0 & 0 \\
0 & 0 & 1 & 0 & \cdots & 0 & 0 \\
\vdots & \vdots & \vdots & \vdots & \vdots & \vdots & \vdots \\
0 & 0 & 0 & 0 & \cdots & 1 & 0 \\
0 & 1 & 0 & 0 & \cdots & 0 & 0 \\
0 & 0 & 0 & 1 & \cdots & 0 & 0 \\
\vdots & \vdots & \vdots & \vdots & \vdots & \vdots & \vdots \\
0 & 0 & 0 & 0 & \cdots & 0 & 1
\end{pmatrix}
\begin{pmatrix}
1 & 0 & 1 & 0 & & & & \\
0 & 1 & 0 & 1 & & & & \\
1 & 0 & -1 & 0 & & & & \\
0 & 1 & 0 & -1 & & & & \\
 & & & & \ddots & & & \\
 & & & & & 1 & 0 & 1 & 0 \\
 & & & & & 0 & 1 & 0 & 1 \\
 & & & & & 1 & 0 & -1 & 0 \\
 & & & & & 0 & 1 & 0 & -1
\end{pmatrix}
$$

$$
\cdot \begin{pmatrix}
1 & 0 & \cdots & 0 & 0 & 0 & \cdots & 0 \\
0 & 0 & \cdots & 0 & 1 & 0 & \cdots & 0 \\
0 & 1 & \cdots & 0 & 0 & 0 & \cdots & 0 \\
0 & 0 & \cdots & 0 & 0 & 1 & \cdots & 0 \\
\vdots & \vdots & \vdots & \vdots & \vdots & \vdots & \vdots & \vdots \\
0 & 0 & \cdots & 1 & 0 & 0 & \cdots & 0 \\
0 & 0 & \cdots & 0 & 0 & 0 & \cdots & 1
\end{pmatrix} \tag{84}
$$

$$
= \begin{pmatrix}
1 & 1 & & & & & & & \\
1 & -1 & & & & & & & \\
 & & 1 & 1 & & & & & \\
 & & 1 & -1 & & & & & \\
 & & & & \ddots & & & & \\
 & & & & & 1 & 1 & & \\
 & & & & & 1 & -1 & & \\
 & & & & & & & 1 & 1 \\
 & & & & & & & 1 & -1
\end{pmatrix}
$$

$$
= B_1
$$

where we can see that premultiplication by P^T and post multiplication by P picks out rows $\left(0, 2, 4, ..., \frac{N}{2} - 1, \frac{N}{2}, \frac{N}{2} + 1, ..., N - 1\right)$ and columns $\left(0, 2, 4, ..., \frac{N}{2} - 1, \frac{N}{2}, \frac{N}{2} + 1, ..., N - 1\right)$ resulting in B_1. Clearly, $P^T(P^T B_2 P) = P^T B_1 = A$ or $\left(P^T\right)^2 B_2 P = A$. We can continue this process and see that $\left(P^T\right)^k B_k P^{k-1} = A$, $k = 1, 2, ..., n - 1$. The last step in the proof is

$$
\begin{aligned}
H_n &= B_n B_{n-1} \cdots B_1 \\
&= \left(B_n P^{n-1}\right)\left(\left(P^T\right)^{n-1} B_{n-1} P^{n-2}\right) \cdots \left(\left(P^T\right)^2 B_2 P\right)\left(P^T B_1\right) .
\end{aligned} \tag{85}
$$

The first term in the above equation, ie, $B_n P^{n-1}$ reduces to A also, which can be seen by noting that $P^{n-1} = P^T$. We have

$$
\begin{aligned}
&\begin{pmatrix}
1 & & & & 1 & & & \\
& 1 & & & & 1 & & \\
& & \ddots & & & & \ddots & \\
& & & 1 & & & & 1 \\
1 & & & & -1 & & & \\
& 1 & & & & -1 & & \\
& & \ddots & & & & \ddots & \\
& & & 1 & & & & -1
\end{pmatrix}
\begin{pmatrix}
1 & 0 & 0 & 0 & \cdots & 0 & 0 \\
0 & 0 & 1 & 0 & \cdots & 0 & 0 \\
\vdots & \vdots & \vdots & \vdots & \vdots & \vdots & \vdots \\
0 & 0 & 0 & 0 & \cdots & 1 & 0 \\
0 & 1 & 0 & 0 & \cdots & 0 & 0 \\
0 & 0 & 0 & 1 & \cdots & 0 & 0 \\
\vdots & \vdots & \vdots & \vdots & \vdots & \vdots & \vdots \\
0 & 0 & 0 & 0 & \cdots & 0 & 1
\end{pmatrix} \\
&= \begin{pmatrix}
1 & 1 & 0 & 0 & \cdots & 0 & 0 \\
0 & 0 & 1 & 1 & \cdots & 0 & 0 \\
\vdots & \vdots & \vdots & \vdots & \vdots & \vdots & \vdots \\
1 & -1 & 0 & 0 & \cdots & & \\
0 & 0 & 0 & 0 & \cdots & 1 & -1
\end{pmatrix} \\
&= A \\
&= B_n P^{n-1} \\
&= B_n P^T
\end{aligned}
\tag{86}
$$

which completes the proof that $H_n = A^n$.

This is an interesting and very useful decomposition as the same structure may be used to transform a data vector, say x, by writing $H_n x = (A(A \cdots (Ax)))$.

This form clearly implies a fast transform requiring work of only $O(n \cdot 2^n)$arithmetic operations. In words, to calculate $H_n x$,

- Construct n columns of 2^n entries each to the right of x,
- Fill in the n columns, left to right, by making the entries in the top half of each column the pairwise sums of the elements in the preceding column, proceeding top to bottom. The entries in the bottom half of each column are the pairwise differences of the elements in the preceding column, proceeding top to bottom.
- The n-th column will contain the Hadamard coefficients.

As an example, consider that

$$x = (1, 4, -2, 3, 0, 1, 4, -1)^T . \tag{87}$$

Direct calculation finds

$$
\begin{aligned}
y &= H_3 x \\
&= (10, -4, 2, -4, 2, -12, 6, 8)^T
\end{aligned}
\tag{88}
$$

and the fast computation proceeds as shown in Table 2.

Since the inverse of H_n is $\frac{1}{2^n}H_n$, we can use the same algorithm for the inverse transform. The only difference is the factor $\frac{1}{2^n}$ which is $\frac{1}{8}$ here.

Table 2 Fast Hadamard Transform Computation

x	Column 1	Column 2	Column 3, y
1	5	6	10
4	1	4	-4
-2	1	-8	2
3	3	4	-4
0	-3	4	2
1	-5	-2	-12
4	-1	2	6
-1	5	-6	8

Chapter 8 Use of the 1–D Hadamard Transform in the Computation of a 2–D Hadamard Transform

Let X be an $N(=2^n) \times M(=2^m)$ dimensional data matrix and let Y be the $N \times M$ Hadamard transform matrix as

$$X = \begin{pmatrix} x_{11} & x_{12} & \cdots & x_{1M} \\ x_{21} & x_{22} & \cdots & x_{2M} \\ \vdots & \vdots & \vdots & \vdots \\ x_{N1} & x_{N2} & \cdots & x_{NM} \end{pmatrix} = \begin{pmatrix} \underline{x}_1 & \underline{x}_2 & \cdots & \underline{x}_M \end{pmatrix} \tag{89}$$

and

$$Y = \begin{pmatrix} y_{11} & y_{12} & \cdots & y_{1M} \\ y_{21} & y_{22} & \cdots & y_{2M} \\ \vdots & \vdots & \vdots & \vdots \\ y_{N1} & y_{N2} & \cdots & y_{NM} \end{pmatrix} = \begin{pmatrix} \underline{y}_1 & \underline{y}_2 & \cdots & \underline{y}_M \end{pmatrix} \tag{90}$$

where $\{\underline{x}_i\}$ and $\left\{\underline{y}_i\right\}$ are N-dimensional column vectors. The 2–D Hadamard transform is given by

$$Y = H_n X H_m \tag{91}$$

where H_n and H_m are $2^n \times 2^n$ and $2^m \times 2^m$ Hadamard matrices. Since n may not necessarily be equal to m, we will use superscripts (n) and (m) to denote the columns $\underline{h}_i^{(n)}$ and $\underline{h}_i^{(m)}$ in H_n and H_m respectively.

A typical entry in Y, say y_{ij}, is given by

$$\begin{aligned} y_{ij} &= \underline{h}_i^{(n)^T} X \underline{h}_j^{(m)} \\ &= \underline{h}_i^{(n)^T} \begin{pmatrix} \underline{x}_1 & \underline{x}_2 & \cdots & \underline{x}_M \end{pmatrix} \begin{pmatrix} h_{1j}^{(m)} \\ h_{2j}^{(m)} \\ \vdots \\ h_{Mj}^{(m)} \end{pmatrix} \\ &= \underline{h}_i^{(n)^T} h_{1j}^{(m)} \underline{x}_1 + \underline{h}_i^{(n)^T} h_{2j}^{(m)} \underline{x}_2 + \cdots + \underline{h}_i^{(n)^T} h_{Mj}^{(m)} \underline{x}_M \end{aligned} \tag{92}$$

or

$$\begin{aligned} y_{ij} &= \left(\left(\underline{h}_i^{(n)^T} h_{1j}^{(m)}\right) \quad \left(\underline{h}_i^{(n)^T} h_{2j}^{(m)}\right) \quad \cdots \quad \left(\underline{h}_i^{(n)^T} h_{Mj}^{(m)}\right) \right) \begin{pmatrix} \underline{x}_1 \\ \underline{x}_2 \\ \vdots \\ \underline{x}_M \end{pmatrix} \\ &= \left(\underline{h}_i^{(n)^T} \otimes \underline{h}_j^{(m)^T}\right) \underline{X} \end{aligned} \tag{93}$$

where $\underline{X}$ is an $N \cdot M$ dimensional vector constructed from the data matrix X in (89). Clearly, a typical column in Y, say $\underline{y}_k$, is given by

$$\underline{y}_k = \begin{pmatrix} \underline{h}_1^{(n)^T} \otimes \underline{h}_k^{(m)^T} \\ \underline{h}_2^{(n)^T} \otimes \underline{h}_k^{(m)^T} \\ \vdots \\ \underline{h}_N^{(n)^T} \otimes \underline{h}_k^{(m)^T} \end{pmatrix} \underline{X} \tag{94}$$

and

$$\begin{aligned} \underline{Y} = \begin{pmatrix} \underline{y}_1 \\ \underline{y}_2 \\ \vdots \\ \underline{y}_M \end{pmatrix} &= (H_n \otimes H_m) \begin{pmatrix} \underline{x}_1 \\ \underline{x}_2 \\ \vdots \\ \underline{x}_N \end{pmatrix} \\ &= H_{n+m} \begin{pmatrix} \underline{x}_1 \\ \underline{x}_2 \\ \vdots \\ \underline{x}_N \end{pmatrix} . \end{aligned} \tag{95}$$

Thus, one $N \times M$ 2-D Hadamard transform can be computed by using a single 2^{n+m} Hadamard transform of an $(N \cdot M)$ data vector constructed from the data matrix.

As an example, let

$$X = \begin{pmatrix} x_{11} & x_{12} & x_{13} & x_{14} \\ x_{21} & x_{22} & x_{23} & x_{24} \end{pmatrix} \tag{96}$$

and

$$Y = \begin{pmatrix} y_{11} & y_{12} & y_{13} & y_{14} \\ y_{21} & y_{22} & y_{23} & y_{24} \end{pmatrix} , \tag{97}$$

then we can write

$$\begin{aligned} Y &= H_1 X H_2 \\ &= \begin{pmatrix} +1 & +1 \\ +1 & -1 \end{pmatrix} \begin{pmatrix} x_{11} & x_{12} & x_{13} & x_{14} \\ x_{21} & x_{22} & x_{23} & x_{24} \end{pmatrix} \begin{pmatrix} +1 & +1 & +1 & +1 \\ +1 & -1 & +1 & -1 \\ +1 & +1 & -1 & -1 \\ +1 & -1 & -1 & +1 \end{pmatrix} , \end{aligned} \tag{98}$$

the elements of which 2×4 matrix may also be computed by

$$\begin{pmatrix} y_{11} \\ y_{21} \\ --- \\ y_{12} \\ y_{22} \\ --- \\ y_{13} \\ y_{23} \\ --- \\ y_{14} \\ y_{24} \end{pmatrix} = \begin{pmatrix} +1 & +1 & +1 & +1 & +1 & +1 & +1 & +1 \\ +1 & -1 & +1 & -1 & +1 & -1 & +1 & -1 \\ +1 & +1 & -1 & -1 & +1 & +1 & -1 & -1 \\ +1 & -1 & -1 & +1 & +1 & -1 & -1 & +1 \\ +1 & +1 & +1 & +1 & -1 & -1 & -1 & -1 \\ +1 & -1 & +1 & -1 & -1 & +1 & -1 & +1 \\ +1 & +1 & -1 & -1 & -1 & -1 & +1 & +1 \\ +1 & -1 & -1 & +1 & -1 & +1 & +1 & -1 \end{pmatrix} \begin{pmatrix} x_{11} \\ x_{21} \\ --- \\ x_{12} \\ x_{22} \\ --- \\ x_{13} \\ x_{23} \\ --- \\ x_{14} \\ x_{24} \end{pmatrix} \tag{99}$$

Chapter 9 The Fast Fourier Transform and the Hadamard Transform

The discrete Fourier transform of a set of data, say $x_0, x_1, ..., x_{N-1}$ is given by the transform coefficients $X_0, X_1, ..., X_{N-1}$ by the relation

$$\begin{pmatrix} X_0 \\ X_1 \\ X_2 \\ \vdots \\ X_{N-1} \end{pmatrix} = \begin{pmatrix} 1 & 1 & 1 & \cdots & 1 \\ 1 & W & W^2 & \cdots & W^{N-1} \\ 1 & W^2 & W^4 & \cdots & W^{2(N-1)} \\ \vdots & & & \vdots & \\ 1 & W^{N-1} & W^{2(N-1)} & \cdots & W^{(N-1)(N-1)} \end{pmatrix} \begin{pmatrix} x_0 \\ x_1 \\ x_2 \\ \vdots \\ x_{N-1} \end{pmatrix} \tag{100}$$

where $W = e^{-j\frac{2\pi}{N}}$. In symbolic form, (100) can be written as

$$\underline{X} = A_N \underline{x} \tag{101}$$

and the inverse transform as

$$\underline{x} = \frac{1}{N} A_N^* \underline{X} \tag{102}$$

where $'*'$ denotes conjugation, and a typical entry, a_{ik}, in A_N is given by

$$a_{ik} = W^{(i-1)(k-1)} \ , \ 1 \le i, k \le N \ . \tag{103}$$

Note that the subscripts in $\underline{x}$ and $\underline{X}$ start with 0. The entries in A_N are on the unit circle, ie, $|a_{ik}| = 1$. Also, $\frac{1}{\sqrt{N}} A_N$ is a unitary matrix.

Direct implementation of (100) requires N^2 multiplications and additions. When N is a power of two, ie, $N = 2^n$, this number can be reduced to $N\,log_2 N$ by the fast Fourier Transform (FFT) algorithm originally derived by Cooley and Tukey (1965). This is, of course, a significant reduction in work.

The FFT is related to the Hadamard transform in some sense. We shall now discuss the basic concepts of this relation. Note that $N = 2^n$ in the following.

Let the data vector be divided into two vectors, with entries having even indices, the $\underline{x}_e$ vector, and the other with odd indices, the $\underline{x}_o$ vector. That is,

$$\begin{aligned} \underline{x}_e^T &= \begin{pmatrix} x_0 & x_2 & x_4 & \cdots & x_{2(\frac{N}{2}-1)} \end{pmatrix} \\ \underline{x}_o^T &= \begin{pmatrix} x_1 & x_3 & x_5 & \cdots & x_{(N-1)} \end{pmatrix} . \end{aligned} \tag{104}$$

We can compute

$$\begin{aligned} \underline{X}_1^T &= (X_0 \quad X_1 \quad X_2 \quad \cdots \quad X_{\frac{N}{2}-1}) \\ \underline{X}_2^T &= (X_{\frac{N}{2}} \quad X_{\frac{N}{2}+1} \quad X_{\frac{N}{2}+2} \quad \cdots \quad X_{N-1}) \end{aligned} \tag{105}$$

separately using the Discrete Fourier Transform (DFT) matrix of order $\frac{N}{2}$, ie,

$$A_{\frac{N}{2}} = \begin{pmatrix} 1 & 1 & \cdots & 1 \\ 1 & W^2 & \cdots & W^{2(\frac{N}{2}-1)} \\ \vdots & \vdots & \vdots & \vdots \\ 1 & W^{2(\frac{N}{2}-1)} & \cdots & W^{2(\frac{N}{2}-1)^2} \end{pmatrix} \tag{106}$$

where

$$\begin{aligned} W^2 &= e^{-j\frac{2\pi}{N}(2)} \\ &= e^{-j\frac{2\pi}{\left(\frac{N}{2}\right)}} \end{aligned} \tag{107}$$

Now,

$$\begin{aligned} \underline{X}_1 = &\begin{pmatrix} X_0 \\ X_1 \\ X_2 \\ \vdots \\ X_{\frac{N}{2}-1} \end{pmatrix} \\ =&A_{\frac{N}{2}} \begin{pmatrix} x_0 \\ x_2 \\ x_4 \\ \vdots \\ x_{2(\frac{N}{2}-1)} \end{pmatrix} \\ &+ \begin{pmatrix} 1 & 0 & 0 & \cdots & 0 \\ 0 & W & 0 & \cdots & 0 \\ 0 & 0 & W^2 & \cdots & 0 \\ \vdots & \vdots & \vdots & \vdots & \vdots \\ 0 & 0 & 0 & \cdots & W^{\frac{N}{2}-1} \end{pmatrix} A_{\frac{N}{2}} \begin{pmatrix} x_1 \\ x_3 \\ x_5 \\ \vdots \\ x_{N-1} \end{pmatrix} \end{aligned} \tag{108}$$

and

$$\begin{aligned} \underline{X}_2 = &\begin{pmatrix} X_{\frac{N}{2}} \\ X_{\frac{N}{2}+1} \\ X_{\frac{N}{2}+2} \\ \vdots \\ X_{N-1} \end{pmatrix} \\ =&A_{\frac{N}{2}} \begin{pmatrix} x_0 \\ x_2 \\ x_4 \\ \vdots \\ x_{2(\frac{N}{2}-1)} \end{pmatrix} \\ &+ \begin{pmatrix} 1 & 0 & 0 & \cdots & 0 \\ 0 & W & 0 & \cdots & 0 \\ 0 & 0 & W^2 & \cdots & 0 \\ \vdots & \vdots & \vdots & \vdots & \vdots \\ 0 & 0 & 0 & \cdots & W^{\frac{N}{2}-1} \end{pmatrix} A_{\frac{N}{2}} \begin{pmatrix} x_1 \\ x_3 \\ x_5 \\ \vdots \\ x_{N-1} \end{pmatrix} \end{aligned} \tag{109}$$

which can be written in symbolic form as

$$\begin{aligned} \underline{X}_1 &= A_{\frac{N}{2}} \underline{x}_e + D_1 A_{\frac{N}{2}} \underline{x}_o \\ \underline{X}_2 &= A_{\frac{N}{2}} \underline{x}_e - D_1 A_{\frac{N}{2}} \underline{x}_o \end{aligned} \tag{110}$$

where the vectors $\underline{x}_e$ and $\underline{x}_o$ are defined in (104) and D_1 is a diagonal matrix with

$$D_1 = dia\left(1, W, W^2, ..., W^{\frac{N}{2}-1}\right) \tag{111}$$

and, as mentioned earlier, $A_{\frac{N}{2}}$ is an $\frac{N}{2}$ order DFT matrix. The two equations in (110) can be put together and can be written in the compact form

$$\begin{aligned} \underline{X} &= \begin{pmatrix} \underline{X}_1 \\ \underline{X}_2 \end{pmatrix} \\ &= \begin{pmatrix} I_{2^{n-1}} & I_{2^{n-1}} \\ I_{2^{n-1}} & -I_{2^{n-1}} \end{pmatrix} \begin{pmatrix} I_{2^{n-1}} & 0 \\ 0 & D_1 \end{pmatrix} \begin{pmatrix} A_{\frac{N}{2}} & 0 \\ 0 & A_{\frac{N}{2}} \end{pmatrix} \begin{pmatrix} \underline{x}_e \\ \underline{x}_o \end{pmatrix} \\ &= B_n(I_{2^{n-1}} \oplus D_1)\left(A_{\frac{N}{2}} \oplus A_{\frac{N}{2}}\right)\begin{pmatrix} \underline{x}_e \\ \underline{x}_o \end{pmatrix} \end{aligned} \tag{112}$$

where B_n is given in (69) in the derivation of the Hadamard transform.

Noting that $A_{\frac{N}{2}}$ is a DFT matrix of order $\frac{N}{2}$, we can take the DFT of $\underline{x}_e$ and $\underline{x}_o$. That is,

$$\begin{aligned} A_{\frac{N}{2}} \underline{x}_e &= C_{n-2}(I_{2^{n-2}} \oplus D_2)\left(A_{\frac{N}{4}} \oplus A_{\frac{N}{4}}\right)\begin{pmatrix} \underline{x}_{ee} \\ \underline{x}_{eo} \end{pmatrix} \\ A_{\frac{N}{2}} \underline{x}_o &= C_{n-2}(I_{2^{n-2}} \oplus D_2)\left(A_{\frac{N}{4}} \oplus A_{\frac{N}{4}}\right)\begin{pmatrix} \underline{x}_{oe} \\ \underline{x}_{oo} \end{pmatrix} \end{aligned} \tag{113}$$

where D_2 can be determined from $A_{\frac{N}{2}}$ in (106), C_n is given in (68), and

$$D_2 = dia\left(1, W^2, W^4, ..., W^{2(\frac{N}{4}-1)}\right) . \tag{114}$$

Note that the number of entries in D_2 is $\frac{N}{4}$ and that the entries are given by W^{2k}, $k = 0, 1, ..., \frac{N}{4} - 1$. The entries in $\underline{x}_{ee}$, $\underline{x}_{eo}$, $\underline{x}_{oe}$, and $\underline{x}_{oo}$ are given by

$$\begin{aligned} \underline{x}_{ee}^T &= (x_0, x_4, x_8, ..., x_{N-4}) \\ \underline{x}_{eo}^T &= (x_2, x_6, x_{10}, ..., x_{N-2}) \\ \underline{x}_{oe}^T &= (x_1, x_5, x_9, ..., x_{N-3}) \\ \underline{x}_{oo}^T &= (x_3, x_7, x_{11}, ..., x_{N-1}) . \end{aligned} \tag{115}$$

The two equations in (113) can be written in the compact form

$$\begin{pmatrix} A_{\frac{N}{2}}\underline{x}_e \\ A_{\frac{N}{2}}\underline{x}_o \end{pmatrix} = B_{n-1}(I_{2^{n-2}} \oplus D_2 \oplus I_{2^{n-2}} \oplus D_2) \cdot \left(A_{\frac{N}{4}} \oplus A_{\frac{N}{4}} \oplus A_{\frac{N}{4}} \oplus A_{\frac{N}{4}}\right)\begin{pmatrix} \underline{x}_{ee} \\ \underline{x}_{eo} \\ \underline{x}_{oe} \\ \underline{x}_{oo} \end{pmatrix} \tag{116}$$

which can be substituted into (112) and we have

$$\underline{X} = B_n(I_{2^{n-1}} \oplus D_1)B_{n-1}(I_{2^{n-2}} \oplus D_2 \oplus I_{2^{n-2}} \oplus D_2) \cdot \left(A_{\frac{N}{4}} \oplus A_{\frac{N}{4}} \oplus A_{\frac{N}{4}} \oplus A_{\frac{N}{4}}\right)\begin{pmatrix} \underline{x}_{ee} \\ \underline{x}_{eo} \\ \underline{x}_{oe} \\ \underline{x}_{oo} \end{pmatrix}. \tag{117}$$

We can continue this procedure resulting in

$$\underline{X} = (B_n D_{11})(B_{n-1}D_{22})(B_{n-2}D_{33})...(B_2 D_{n-1,n-1})B_1(P_b\underline{x}) \tag{118}$$

where

$$\begin{aligned} D_{11} &= (I_{2^{n-1}} \oplus D_1) \\ D_{22} &= (I_{2^{n-2}} \oplus D_2 \oplus I_{2^{n-2}} \oplus D_2) \\ &\vdots \\ D_{n-1,n-1} &= (I_2 \oplus D_{n-1} \oplus ... \oplus I_2 \oplus D_{n-1}) \end{aligned} \tag{119}$$

with

$$\begin{aligned} D_1 &= dia\left(1, W, ..., W^{\frac{N}{2}-1}\right) \\ D_2 &= dia\left(1, W^2, ..., W^{2(\frac{N}{4}-1)}\right) \\ D_3 &= dia\left(1, W^4, ..., W^{4(\frac{N}{8}-1)}\right) \\ &\vdots \\ D_{n-1} &= dia\left(1, W^{\frac{N}{4}}\right) \end{aligned} \tag{120}$$

the $B_j's$ are given in (69), and $P_b\underline{x}$ is a permuted data vector. P_b is sometimes referred to as a 'bit reversal matrix.' This matrix appears because the data is shuffled at each stage in the algorithm development. See for example (115) — (116). This can be implemented by the following procedure:

i. Identify the data location, say $\underline{l}'$.
ii. Write it in binary form $\underline{l}' = \left(\underline{l}'_{n-1}, \underline{l}'_{n-2}, ..., \underline{l}'_0\right)$ where we have assumed that $N = 2^n$.
iii. Define $\underline{l}'_{rev}$ as the reversed order of bits, ie, $\underline{l}'_{rev} = \left(\underline{l}'_0, \underline{l}'_1, ..., \underline{l}'_{n-1}\right)$.
iv. Find the decimal equivalent of $\underline{l}'_{rev}$.

For example, for $N = 16$ and $\underline{l}' = 11 = (1, 0, 1, 1)_2$ then $\underline{l}'_{rev} = (1, 1, 0, 1)_2 = 13$. $\underline{x}^T$ and $P_b\underline{x}^T$ are given by
$\underline{x}^T = \left(x_0 x_1 x_2 x_3 x_4 x_5 x_6 x_7 x_8 x_9 x_{10} \overbrace{x_{11}} x_{12} x_{13} x_{14} x_{15}\right)$ and
$(P_b\underline{x})^T = \left(x_0 x_8 x_4 x_{12} x_2 x_{10} x_6 x_{14} x_1 x_9 x_5 \overbrace{x_{13}} x_3 x_{11} x_7 x_{15}\right)$ where the overbraces identify the entries referred to above.

The algorithm above, ie, the implementation of (100) by (118) is usually referred to as a fast Fourier transform (FFT) of N point data with decimation in time, as the algorithm is structured to appeal to the concept of sample rate reduction. Clearly, A_N in (101) can be expressed as

$$A_N = (B_n D_{11})(B_{n-1} D_{22})(B_{n-2} D_{33}) \cdots (B_2 D_{n-1,n-1}) B_1 P_b \,. \qquad (121)$$

Since the DFT matrix A_N is symmetric, $A_N = A_N^T$, $B_i^T = B_i$, and $D_{ii}^T = D_{ii}$, it follows that

$$A_N = P_b^T B_1 (D_{n-1,n-1} B_2) \cdots (D_{33} B_{n-2})(D_{22} B_{n-1})(D_{11} B_n) \qquad (122)$$

and the resultant algorithm is usually referred to as the FFT of an N point data set with decimation in frequency. Using the results in (76), ie, using the matrix A, we can write

$$\begin{aligned} A_N &= A(P_{n-1}^T D_{11} P_{n-1}) A(P_{n-2}^T D_{22} P_{n-2}) \cdots A(P_1^T D_{n-1,n-1} P_1) A P_b \\ &= A D'_{11} A D'_{22} \cdots D'_{n-1,n-1} A P_b \end{aligned} \qquad (123)$$

where $D'_{ii} = P_{n-i}^T D_{ii} P_{n-i}$ are diagonal matrices with complex entries. Clearly, the differences in the Hadamard matrix transform matrix expansion in terms of a power of the 'A' matrix, ie, $H_n = A^n$ and the DFT matrix decomposition in (123) are the diagonal matrices $\left\{D'_{ii}\right\}$ and the bit reversal matrix, P_b.

We have presented three different algorithms that have strong ties to the Hadamard transform in that the underlying structures, or some modifications of them, are the B_i matrices. We note that there are other algorithms and the interested reader is referred to Oppenheim and Schafer (1989).

Chapter 10 Computational Structures

There are a number of structures which can be constructed to compute the fast Sylvester-Hadamard transform. Some of these structure are directly useful in designing a hardware implementation and some are useful for a deeper insight to the transform.

Vetterli (1983) discloses a binary delay tree which is an excellent example of a decomposition into orthogonal subblocks. The orthogonality of subblocks is a necessity (and sufficiency) for an *orthogonal* transform. The basic building block of such a tree is shown in Figure 1.

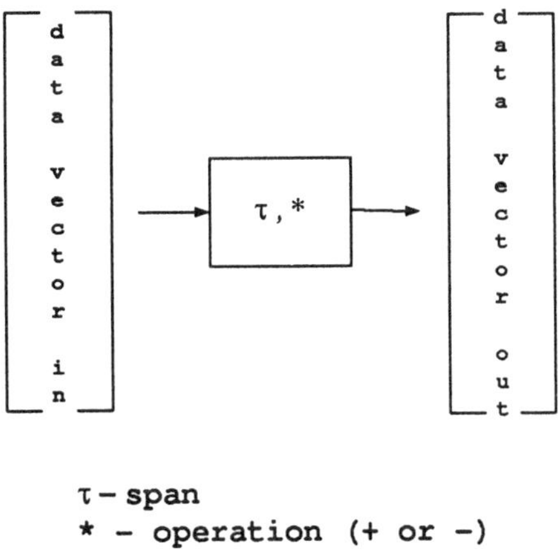

Figure 1 The Basic Building Block of a Binary Delay Tree

The binary delay tree basic building block has two parameters. The first, $'\tau'$, is the span of binary computation and the second, '*', is the computational operation; here, either addition or subtraction. The block operates on an input data vector, proceeding from the top, and producing an output data vector whose components, also proceeding from the top, are pairwise summations or differences of the elements in the input data vector which are separated by a span of τ. As an example, we use Vetterli's construction to calculate H_3X in Figure 2. Note that H_3X appears in Sylvester ordering, top to bottom, at the extreme right of the tree.

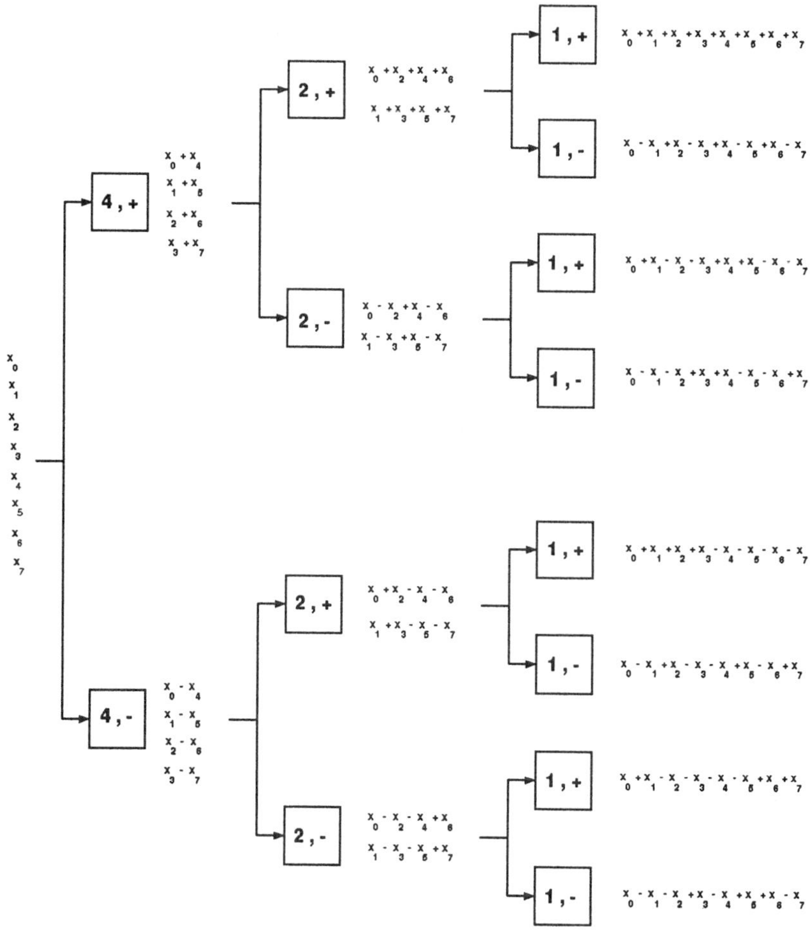

Figure 2 Computation of H_3X using Vetterli's Binary Delay Tree

Vetterli's binary tree could be directly realized in hardware. It does not, however, distribute the computational load evenly across the computational building blocks and is therefore not a *systolic* computational architecture as discussed in Kung (1984) or Hassan et al (1994). For applications where a systolic attribute is desirable, the reader is referred to Lee and Yasuda (1990).

Finally, the reader is referred to a very clever implementation of the Fourier transform via a modular Hadamard structure. This implementation is termed the "Green Machine" and was used to decode telemetry from Mariner '69 (Mann, ed., 1969).

Chapter 11 Efficient Hadamard Transform Domain Computation

The Hadamard domain is well suited for certain calculations of special interest to pattern analysis. Pearl (1971) looks at the Hadamard domain in light of his recognition that "... *the merit of every mathematical tool is in the simplicity and ease at which it facilitates calculations for a given class of problems* ... ." One of these problem classes is the calculation of the average Hamming distance between two n-bit independent, randomly generated vectors.

For motivation, let us assume that n=3 and that the two patterns, $\mathcal{A}$ and $\mathcal{B}$ have probability density functions (pdf's), $P_{\mathcal{A}}$ and $P_{\mathcal{B}}$, and respective transforms, $P^*_{\mathcal{A}}$ and $P^*_{\mathcal{B}}$, as set forth in Table 3.

Table 3 Probability Density Functions and Their Transforms

Pattern			$P_{\mathcal{A}}$	$P^*_{\mathcal{A}}$	$P_{\mathcal{B}}$	$P^*_{\mathcal{B}}$
0	0	0	1/4	1	1/8	1
0	0	1	0	-1/4	3/8	-1/4
0	1	0	1/8	0	1/8	1/2
0	1	1	1/4	1/4	0	-1/4
1	0	0	0	1/4	1/8	1/4
1	0	1	1/4	1/2	1/8	0
1	1	0	0	-1/4	0	1/4
1	1	1	1/8	1/2	1/8	-1/2

If we wished to calculate the average Hamming distance, $\overline{D_{\mathcal{AB}}}$, between an $\mathcal{A}$—vector pattern and a $\mathcal{B}$—vector pattern, we could straightforwardly sum the appropriately weighted components of the $2^n \cdot 2^n = 2^{2n}$ possible products. For the case of Table 3 this direct approach yields $\overline{D_{\mathcal{AB}}} = 23/16$.

Pearl (1971) shows, however, that this problem can be solved with far less computation by first taking the Hadamard transforms of the pdfs, and then summing only n products. The key is in noting that the transform of the pdf of a random variable, $\mathcal{C}$, which is the bit-by-bit modulo-two sum of vectors $\mathcal{A}$ and $\mathcal{B}$, is

$$P^*_{\mathcal{C}}(k) = P^*_{\mathcal{A}}(k)P^*_{\mathcal{B}}(k)\,,\ 0 \le k < 2^n - 1\,. \tag{124}$$

Pearl (1971) then goes on to show that $\overline{D_{\mathcal{AB}}}$ can be calculated using only those coefficients of the pdf transforms that have indices of powers of two, ie,

$$\overline{D_{\mathcal{AB}}} = \frac{n}{2} - \frac{1}{2}\sum_{i=0}^{n-1} P^*_{\mathcal{A}}(2^i)P^*_{\mathcal{B}}(2^i)\,. \tag{125}$$

As the reader can easily verify, the form (125) indeed also yields 23/16 when used on the sample problem.

Chapter 12 The Hadamard Transform in Statistics

Use of the Hadamard transform, and the devisement of a fast algorithm for its implementation, goes back to Yates (1937) who used the transform to determine what factors were contributory in a factorial experiment, ie, to uncover the dependencies within a multivariable data set.

As motivation, consider the following *discrete* problem which can be found in Stoyanov (1987) and Buehler et al (1981) and is a specific case of an n-variable general construction given in Driscoll (1978). Suppose we have three memoryless binary variables, X, Y, and Z. Suppose that we estimate, by histogramming for example, the pdf of the triples (X,Y,Z) as depicted in Figure 3.

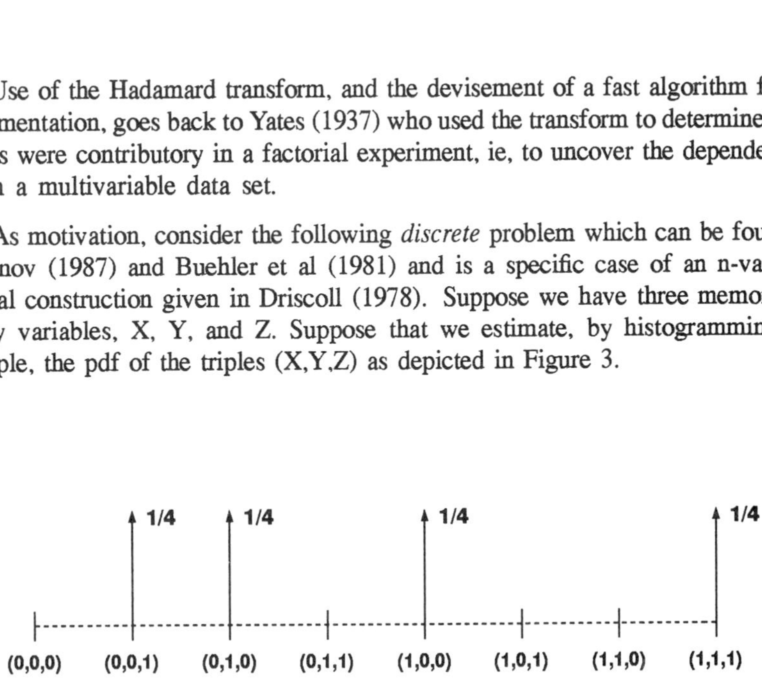

Figure 3 The pdf of the Triple (X,Y,Z)

It is clear from Figure 3 that the three variables X, Y, and Z are mutually dependent as only four out of the eight possible triples occur. Suppose, however, that we had looked only at the marginal pdf's or at the pdf's for the variables taken two at a time. These pdf's would have demonstrated that the individual variables appeared to take on the values 0 and 1 with equal probability and that each variable pair took on the values (0,0), (0,1), (1,0), and (1,1) with equal probability. We would not have concluded dependence until we examined the triples.

The Hadamard transform as used by Yates (1937) is an excellent efficient tool for sniffing out the dependence among a set of variables. Consider again that we have three binary variables X, Y, and Z. The variables might indicate the presence or absence (1 or 0) of a particular factor in a an agricultural experiment or it might indicate the presence or absence of a particular feature in a pattern recognition experiment. Let us assume that an experiment is tried 80 times with the results as summarized in Figure 4.

X	Y	Z	Data	I	II	III	
0	0	0	12	21	42	80	
0	0	1	9	21	38	-2	
0	1	0	8	18	-2	-2	
0	1	1	13	20	0	16	←
1	0	0	11	3	0	4	
1	0	1	7	-5	-2	-2	
1	1	0	8	4	8	2	
1	1	1	12	-4	8	8	

Figure 4 Hadamard Transform of a Specific Data Set of 80 Values with the Outlier Coefficient Indicated by Arrow

The data exhibits a χ^2 value of 3.6. There are 7 degrees of freedom and thus there is small motivation for rejecting the null hypothesis of no dependency among the variables X, Y, and Z. The Hadamard transform discloses a different picture, however. The first entry in the Hadamard transform (Column III) is, of course, the total size of 80 as this entry corresponds to the all ones row of H_3. The other coefficients should hover about zero if there is indeed no dependence among the variables. One coefficient, 16, stands out as a significant departure from zero for a sample size of 80. To interpret this we look at the corresponding row of the factorial enumeration or *truth table* of the variables X, Y, and Z. The row of the truth table corresponding to the outlier variable is 0,1,1. This is to be interpreted as disclosing a dependence among the variables that are ones in the truth table entry, in this case, Y and Z. That there is such a dependence is clear. A quick way to see it in this case is to consider that if the data were random, the expected value or mean of each entry in the Data column would be 80/8=10. For those rows where Y and Z are the same, the value in the Data column is greater than the mean; for those rows where Y and Z are different, the value is less than the mean.

Thus, the Hadamard transform, a linear statistical test, may be far more useful as a statistical diagnostic tool than the far less complex, and more easily and quickly implemented, nonlinear χ^2-test.

Chapter 13 The Hadamard Transform and Error-Correction Coding

Harwit and Sloane (1979) point out some connections between Hadamard matrices and error correcting codes. In particular they address the code constructed from H_n created by augmenting H_n with its inverse, ie, $-H_n$ and subsequently changing the ones to zeros and the minus ones to ones. The set of 2^{n+1} 2^n-bit words forms a code which has minimum distance 2^{n-1}. They remark that this forms a first order Reed-Muller code.[3]

As an example, Harwit and Sloane (1979) show that the code formed from H_3 consists of the 16 8-bit words listed in Table 4.

Table 4 A Sixteen Word Single Error Correcting Code

CODE WORD NUMBER	CODE WORD
0	0 0 0 0 0 0 0 0
1	0 1 0 1 0 1 0 1
2	0 0 1 1 0 0 1 1
3	0 1 1 0 0 1 1 0
4	0 0 0 0 1 1 1 1
5	0 1 0 1 1 0 1 0
6	0 0 1 1 1 1 0 0
7	0 1 1 0 1 0 0 1
8	1 1 1 1 1 1 1 1
9	1 0 1 0 1 0 1 0
10	1 1 0 0 1 1 0 0
11	1 0 0 1 1 0 0 1
12	1 1 1 1 0 0 0 0
13	1 0 1 0 0 1 0 1
14	1 1 0 0 0 0 1 1
15	1 0 0 1 0 1 1 0

[3] A Reed-Muller code of order r is defined for code words of length 2^n having k-information bits where $k = 1 + \binom{n}{1} + \cdots + \binom{n}{r}$ (Peterson and Weldon, 1972).

The code can be decoded using the Hadamard transform. We first convert the zeros to plus ones and the ones to minus ones. Then we premultiply by H_n and locate the largest magnitude coefficient of the transform. Let this coefficient be found at the j-th row. If this coefficient is positive then we declare that the j-th code word was sent; if negative, we declare for the $j + 2^n$-st code word. As an example, we pick the 5-th code word, ie, 0 1 0 1 1 0 1 0. We convert it to +1 -1 +1 -1 -1 +1 -1 +1 and take its Hadamard transform we obtain

$$H_3 \begin{pmatrix} +1 \\ -1 \\ +1 \\ -1 \\ -1 \\ +1 \\ -1 \\ +1 \end{pmatrix} = \begin{pmatrix} 0 \\ 0 \\ 0 \\ 0 \\ 0 \\ +8 \\ 0 \\ 0 \end{pmatrix}. \tag{126}$$

If we had selected the 13-th code word, we would have found a –8. Now note that any *single* error will change the magnitude of *each* of the transform coefficients by 2.

Many communication environments are extremely noisy. The noise mechanism may be quite complex such as a non stationary mix of impulsive processes, Gaussian noise, and frequency selective fading. Such "unfriendly channels" will become ever more commonplace as communication continues to be impressed into use in challenging environments such as in factories and around other heavy machinery. It is often not possible to use familiar and efficient error control techniques against the high error rates encountered. The good news is that the amount of information that must be conveyed through a very noisy environment is often nominal and the encoding need be robust rather than efficient.

In many cases, repetition coding is used as a stand-alone encoding or as an inner code. Repetition coding is extremely inefficient and we can do better. One way is to use a new very low rate coding method, *Random Parity Coding*, (RPC), which may be used in the transport of information over very noisy channels. RPC was introduced by Hershey and Tiemann (1996) and may be used either as a stand-alone encoding scheme or as a method of reducing the error rate so that a powerful and efficient forward error correction code can be used in conjunction with it.

The encoding method and apparatus is keyed to Figure 5. The encoder consists of a k-bit information word $(c_1, c_2, \ldots, c_k)$ that is "dot-producted" against a succession of k-bit "projectors," which are k-bit words chosen at random or generated pseudorandomly.[4] Let n be the number of projectors. These words, and their sequence, is also assumed to be known by the decoder. The parity of the successive dot products is sent as the output bits of the coder.

[4] A projector is not allowed to be all zeros as this case would convey no information respecting the information word.

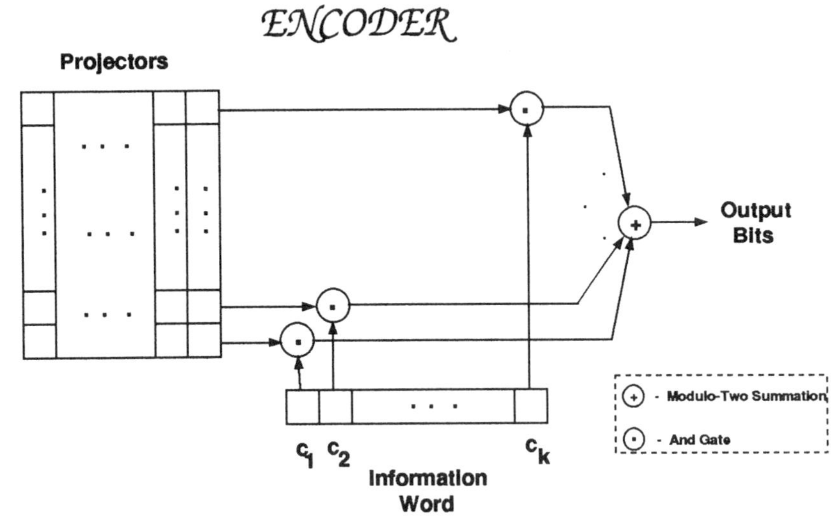

Figure 5 Random Parity Encoder

The decoder is depicted in Figure 6.

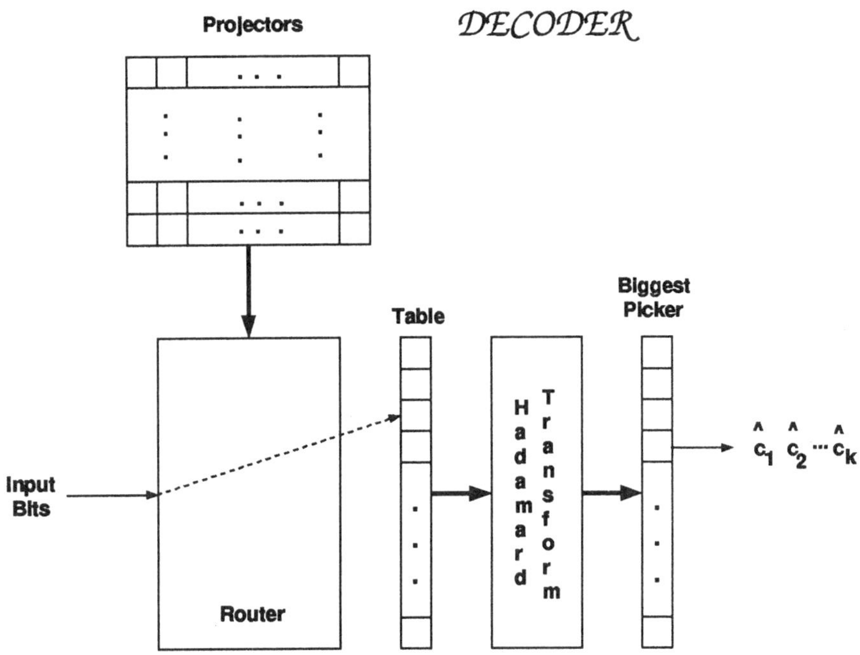

Figure 6 Random Parity Decoder

The decoder receives the incoming bits which have been generated at the encoder and subsequently passed through a noisy channel. In the decoder, the projectors serve as successive k-bit addresses to a table. If the incoming i-th bit is a zero, the content of the indicated table location is incremented by w_i; if the incoming bit is a one, the content is decremented by w_i. The increment/decrement, w_i, is assumed here to be unity. (It need not be so constrained as the code can also be soft decoded with side information and this may result in a different weighting for each i.)

After the table has been filled, the table contents are considered as a $2^k \times 1$ vector, V. The Hadamard transform is then taken. This is done by forming $H_k V$. The biggest (most positive) coefficient of the transform is then identified. The table *address* corresponding to the location of the biggest coefficient is the estimate of the decoded information word, $(\widehat{c_1}, \widehat{c_2}, ..., \widehat{c_k})$.

As an example, Table 5 depicts the action of the Random Parity Encoder for n=22 and k=3 where $(c_1, c_2, c_3) = (1, 1, 0)$. The code word produced is

Table 5 Encoding Example

RANDOMLY CHOSEN PROJECTORS			OUTPUT BIT
1	1	1	0
0	0	1	0
0	0	1	0
1	0	0	1
1	0	0	1
0	1	1	1
1	1	1	0
1	1	0	0
0	0	1	0
0	1	0	1
0	0	1	0
1	0	0	1
1	0	1	1
1	0	0	1
0	1	0	1
1	0	0	1
1	0	0	1
1	0	1	1
0	1	1	1
0	0	1	0
0	1	0	1
1	1	1	0

0001110001011111111010 as shown in the Output Bit Column. In the Random Parity Decoder, the projectors designate the table locations which are to be incremented/decremented by unity if the received bit is a zero/one. Table 6 presents the projectors, the table addresses specified by them, and the appropriate action, increment (inc), or decrement (dec), as determined by the received bits.

Table 6 Decoding Example

PROJECTORS			ADDRESS/ACTION
1	1	1	7/inc
0	0	1	1/inc
0	0	1	1/inc
1	0	0	4/dec
1	0	0	4/dec
0	1	1	3/dec
1	1	1	7/inc
1	1	0	6/inc
0	0	1	1/inc
0	1	0	2/dec
0	0	1	1/inc
1	0	0	4/dec
1	0	1	5/dec
1	0	0	4/dec
0	1	0	2/dec
1	0	0	4/dec
1	0	0	4/dec
1	0	1	5/dec
0	1	1	3/dec
0	0	1	1/inc
0	1	0	2/dec
1	1	1	7/inc

The decoder table is filled according to the address/action column using the corresponding projectors as address. Thus, for example, location 1 of the table is set to 5 as there are 5 occurrences of 1/inc and no occurrences of 1/dec. The decoder table state subsequent to receiving and processing all of the twenty-two bits is depicted in Table 7.

Table 7 Decoder Table Contents after Receiving and Processing the 22–bit Code Word

TABLE ADDRESS	CONTENTS
0	0
1	+5
2	-3
3	-2
4	-6
5	-2
6	+1
7	+3

The Hadamard transform is taken of the 8x1 Contents Vector, V, ie, we compute

$$H_3 V = \begin{pmatrix} -4 \\ -12 \\ -2 \\ -6 \\ +4 \\ 0 \\ +22 \\ -2 \end{pmatrix}, \tag{127}$$

and determine that its largest coefficient is +22. This corresponds to table entry 6 which has a binary equivalent of 1 1 0 in binary and is our estimate of the transmitted code word.

Let us now examine the effect of channel errors on the decoding process. Let us assume that five bits out of the twenty two sent were received incorrectly. Let these bits be identified in boldface as indicated in Table 8.

Table 8 Transmission Errors

No Errors	0 0 0 1 1 1 0 0 0 1 0 1 1 1 1 1 1 1 1 0 1 0
Errors	0 **1** 0 1 1 1 **1** **1** 0 1 0 1 **0** 1 1 1 1 1 **0** 0 1 0

The effects of the five transmission errors are reflected in the Decoder Table Contents as depicted in Table 9.

Table 9 Decoder Table Contents after Receiving and Processing the 22-bit Code Word Affected by 5 Transmissions Errors

TABLE ADDRESS	CONTENTS
0	0
1	+3
2	-3
3	0
4	-6
5	0
6	-1
7	+1

The errors change the Hadamard transform coefficients and we find that

$$H_3V = \begin{pmatrix} -6 \\ -14 \\ 0 \\ -4 \\ +6 \\ +2 \\ +12 \\ +4 \end{pmatrix} . \tag{128}$$

The largest coefficient of H_3V is the value +12. Notice that the largest coefficient still identifies the correct code word although the coefficient's value has been reduced by 10 *which is a value equal to twice the number of errors in our example.*

Proceeding on, we select the general Binary Symmetric Channel (BSC) model as our data transport channel model. This channel model is memoryless and transports a bit correctly with probability q and incorrectly with probability $p = 1 - q$. The capacity, C_p, of this channel is well known to be

$$C_p = 1 + plog_2p + qlog_2q \, . \tag{129}$$

Figure 7 displays the results in using RPC for three values of k. These results are contrasted with the results obtained with simple repetition coding. The vertical dotted line is the "Shannon wall," which is the inverse of the capacity of the BSC. Note that the x-axis can be interpreted as the reciprocal of the code rate or as the number of BSC uses per information bit.

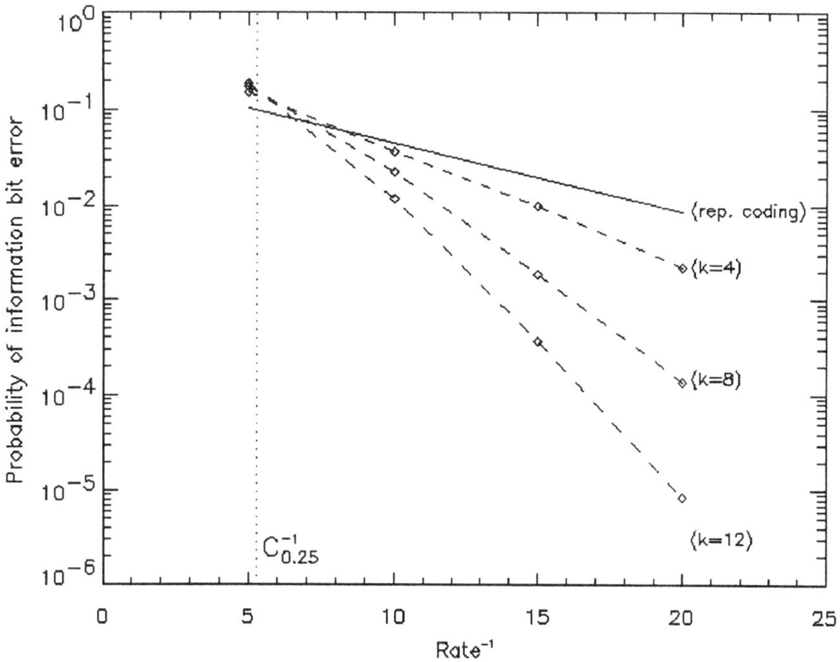

Figure 7 Simulated Performance of RPC for k=4, 8, and 12 over a BSC with p=0.25

We can develop an approximation to the error probability as a function of n, the number of projectors, and k, the number of information bits in the code word.

To do this, we recognize that the Hadamard transform coefficients are of two classes:

- Class A — the coefficient which properly identifies the k-bit information word. There is 1 member in class A.
- Class B — the coefficients which incorrectly identify the k-bit information word. There are $2^k - 1$ members in class B.

The coefficient in Class A is designated as x_a and is modeled as

$$x_a \in 2B(n, q) - n \tag{130}$$

where B(n,q) is a binomial variate, ie, a sum of n independent Bernoulli trials each with probability q. A coefficient in Class B is designated as x_b and is modeled as

$$x_b \in 2B\left(n, \frac{1}{2}\right) - n\,. \tag{131}$$

The class A member must exceed all of the class B members in order to achieve error-free decoding. The probability that the class A coefficient exceeds any particular class B member is given by the probability that $y > 0$ where y is defined as

$$y = x_a - x_b \,. \tag{132}$$

Approximating the binomial variates as Gaussians, $G(\mu, \sigma^2)$, of mean μ and variance σ^2, we have

$$\begin{aligned} x_a &\in G(n(q-p), 4npq) \\ x_b &\in G(0, n) \end{aligned} \tag{133}$$

and consequently

$$y \epsilon\, G(n(q-p), n(1+4pq)) \,. \tag{134}$$

Defining δ as the ratio of the mean of y to its standard deviation, we have

$$\delta = \sqrt{n} \frac{q-p}{\sqrt{1+4pq}} \tag{135}$$

and we may then define P_c as the probability that $y > 0$ for any particular class B member and write

$$P_c = 1 - Q(\delta) \tag{136}$$

where

$$Q(\delta) \equiv \frac{1}{\sqrt{2\pi}} \int_{\delta}^{\infty} e^{-\frac{t^2}{2}} dt \,. \tag{137}$$

The probability of correct decoding, P_{CD}, is approximated by assuming all of the above and independence between the Hadamard coefficients. We then have

$$P_{CD} = P_c^{2^k - 1} \,. \tag{138}$$

Next, we investigate the probability of decoded bit error versus α at a particular value of p.

For large δ,

$$Q(\delta) \approx \frac{1}{\sqrt{2\pi}\delta} e^{-\frac{\delta^2}{2}} \tag{139}$$

and

$$P_{CD} \approx P_c^{2^k} \approx \left(1 - \frac{1}{\sqrt{2\pi}\delta} e^{-\frac{\delta^2}{2}}\right)^{2^k} \,. \tag{140}$$

Recalling that the definition of the base of natural logarithms, e, is

$$e = \lim_{h \to 0} (1+h)^{\frac{1}{h}} \tag{141}$$

where h may approach zero from either side, we set

$$h = -\frac{1}{\sqrt{2\pi}\delta} e^{-\frac{\delta^2}{2}} \tag{142}$$

in (140) and we then have

$$P_{CD} \approx e^{2^k h} . \tag{143}$$

The exponent, $2^k h$ in (143) is critical. If it goes to zero as k becomes increasingly large, then correct decoding will obtain.

Writing

$$2^k h \approx -e^{\left(-\frac{\delta^2}{2} + k \ln 2\right)} \tag{144}$$

for large δ, we further define z as

$$z = -\frac{\delta^2}{2} + k \ln 2 . \tag{145}$$

Letting ρ be defined as

$$\rho = \frac{q-p}{\sqrt{1+4pq}} \tag{146}$$

and let us express n as

$$n = \frac{\alpha k}{C_p} . \tag{147}$$

We can then write z as

$$z = -\frac{\alpha n \rho^2}{2C_p} + k \ln 2 \tag{148}$$

or

$$z = \left(\ln 2 - \frac{\alpha \rho^2}{2C_p} \right) k . \tag{149}$$

For $P_{CD} \rightarrow 1$ it is necessary that $z \rightarrow -\infty$. This will be the case as $k \rightarrow \infty$ if

$$\alpha > \frac{2C_p ln2}{\rho^2}. \tag{150}$$

Defining $\alpha_0 \equiv \frac{2C_p ln2}{\rho^2}$, we note that $lim\, \alpha_0 = 2$ as $p \rightarrow 1/2$. We plot α_0 in Figure 8 and superimpose the ratio, C_p/R_0, of the BSC's capacity to the BSC's cutoff rate, R_0, where

$$R_0 = 1 - log_2(1 + 2\sqrt{pq}). \tag{151}$$

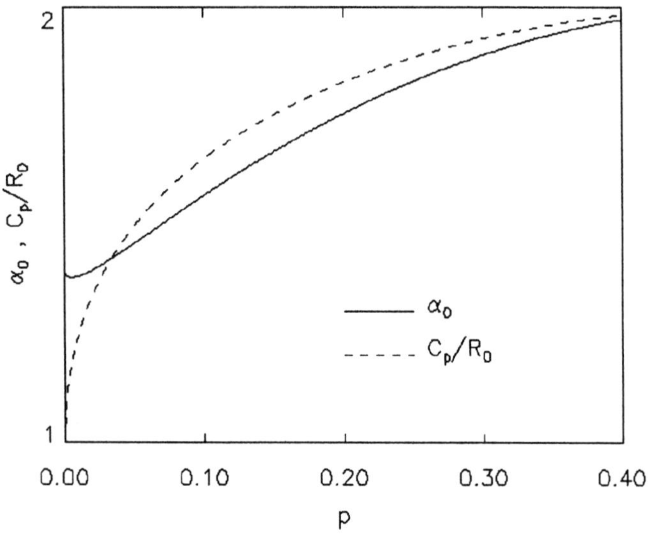

Figure 8 α_0 and C_p/R_0 as a Function of p

Chapter 14 Boolean Functions

In conjunction with Hadamard matrices, boolean functions can be thought of as (1) representations of patterns, or, (2) lossy multiplexers or data combiners. For this discussion, let f_d denote a boolean function of degree d defined as a mapping from the 2^d states $(x_1, x_2, ..., x_d)$, $x_i \in \{0, 1\}$, such that $f_d(x_1, x_2, ..., x_d) \in \{\pm 1\}$.

We may define the Hadamard transform of a boolean function as per Rothaus (1976). Let us set

$$\underline{\alpha} = H_d f_d = \begin{pmatrix} \alpha_0 \\ \alpha_1 \\ \vdots \\ \alpha_{2^d-1} \end{pmatrix}. \tag{152}$$

Because the Hadamard transform is a true Fourier Transform, sometimes called the *BIFORE* transform, Rothaus observes that Parseval's relation obtains and that we have $\sum\limits_{i} \alpha_i^2 = 2^{2d}$. Hershey (1980) concluded that although the Parseval relationship implies a dependency between the $\{\alpha_i\}$, it is reasonable to consider them independent for a boolean function of nominal degree and it is reasonable to expect that there will be some relatively large magnitude $\alpha_i's$ for a boolean function or pattern chosen at random.

Chapter 15 Spectrally Preconditioned Threshold Logic

In a more visual sense, consider f_d to be a pattern consisting of a pasting of black and white stickers on some or all of the vertices of a d-dimensional boolean hypercube. It is at this early juncture that two important questions arise. First, how can we efficiently code or store a random pattern, or how, when presented with a d-bit address of a vertex on the hypercube, can we decide, either with total certainty or within a given error tolerance, the color at the vertex without using a look up technique? Second, some patterns appear more random or complex than others. Can this random quality be defined and is there a metric or even a fuzzy measure for it?

Early attempts to deal with these questions involved threshold logic, specifically the *Threshold Logic Unit*, or TLU. The first milestone of this research was the creation of the PERCEPTRON (Rosenblatt, 1962) which was an algorithm and implementing device which learned to separate or dichotomize linearly separable patterns.

A linearly separable boolean function of d-variables is one for which there exists a set of weights, $\{w_1, w_2, ..., w_d, \Theta\}$ such that $\sum_{i=1}^{d} w_i x_i + \Theta > 0$ if, and only if, $f_d(x_1, x_2, ..., x_d) = 1$ and $\sum_{i=1}^{d} w_i x_i + \Theta < 0$ if, and only if, $f_d(x_1, x_2, ..., x_d) = -1$. (The case in which $\sum_{i=1}^{d} w_i x_i + \Theta = 0$ is considered to be *unresolved*.) The TLU is depicted in Figure 9. The inputs are denoted by $\{x_i\}$ and the output by y. The inputs are summed by the Σ element and added to Θ. If this sum is equal to or greater than zero, the output is a $+1$, if less than zero it is a -1. (If equal to zero, the *unresolved* case, the output is considered to be zero.)

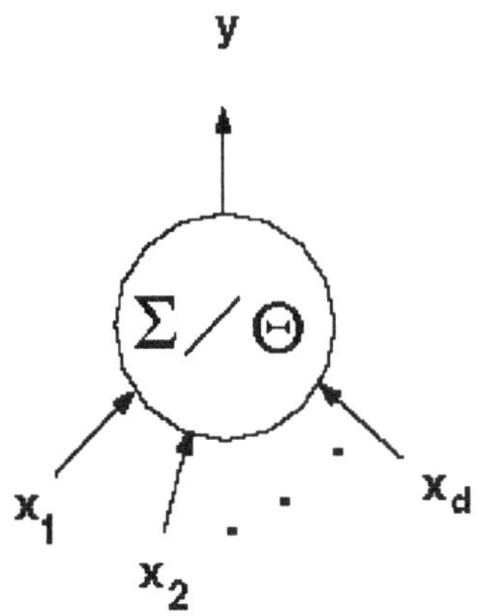

Figure 9 The Threshold Learning Element

The remarkable thing is that there is an algorithm which, in a finite number of iterations, will determine a set of weights separating a linearly separable function.[5] The algorithm operates in a training mode. The weights and threshold $(w_1, w_2, ..., w_d, \Theta)$ are set to an initial value and the following iteration effected: $(w_1, w_2, ..., w_d, \Theta) \cdot (x_1, x_2, ..., x_d, 1)^T$ is computed for each pattern point. (Remember that the $\{x_i\}$ are zeros and ones here.) If the inner product is of the same sign as $f_d(x_1, x_2, ..., x_d)$ then no change or *correction* is made to $(w_1, w_2, ..., w_d, \Theta)$. If, however, the signs are not the same, then $(w_1, w_2, ..., w_d, \Theta) \leftarrow cf_d(x_1, x_2, ..., x_d)(w_1, w_2, ..., w_d, \Theta)$ is performed for this particular pattern point, ie, each weight or the threshold is replaced by its previous value times the scalar, c times the value of f_d. The scalar multiplier, $c > 0$, is the correction increment and it may be of arbitrary magnitude. In practice it is often chosen to be unity.

The entire set of 2^d is cycled through again and again until no corrections are made at any of the points. The algorithm is then declared successfully terminated and the solution is the final value of $(w_1, w_2, ..., w_d, \Theta)$. If the function or pattern is not linearly separable, then application of this training procedure will never terminate.

There are, of course, 2^{2^d} boolean functions of d-variables, but, for even nominal d, only a small fraction of these functions are linearly separable. Hu (1965) tabulates the number of such functions. Table 10[6] lists the total number of boolean functions for the range $1 \leq d \leq 6$, the corresponding number of linearly separable functions, and the percentage of these separable functions to the total number of boolean functions of the same degree.

Table 10 Enumeration of Threshold Boolean Functions up to Degree Six

Degree	Number of Boolean Functions	Number of Separable Functions	Percentage of Separable Functions
1	4	4	100
2	16	14	87.5
3	256	104	40.6
4	65536	1882	2.9
5	$4.3 \cdot 10^9$	94572	$2.2 \cdot 10^{-3}$
6	$1.8 \cdot 10^{19}$	15028134	$8.3 \cdot 10^{-11}$

Hershey (1980) proposed a variant to the TLU wherein a non linearly separable pattern may be rendered separable by remapping it into a coordinate system whose axes correspond to the largest Hadamard transform coefficients as discussed in

5 An excellent treatment is provided by Nilsson (1965).

6 This table is prepared according to a style suggested by Carne (1965).

Chapter 12, *The Hadamard Transform in Statistics*. As an example, consider the three variable boolean function, f_3, and its Hadamard transform given in Table 11.

Table 11 The Example Three Variable Boolean Function f_3

x_1	x_2	x_3	$f_3(x_1, x_2, x_3)$	$H_3 f_3$
0	0	0	+1	0
0	0	1	-1	+4
0	1	0	+1	0
0	1	1	-1	-4
1	0	0	-1	0
1	0	1	+1	+4
1	1	0	+1	0
1	1	1	-1	+4

For sake of clarity, we represent a +1 by an open circle and a −1 by a shaded circle. In three dimensions, f_3 appears as shown in Figure 10.

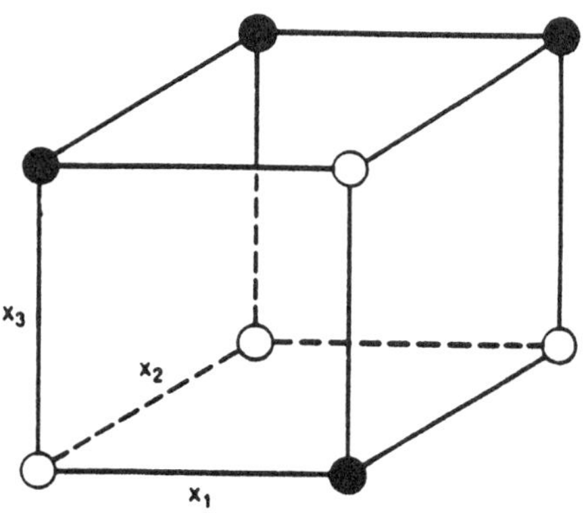

Figure 10 f_3

Clearly, there is no linear surface (plane) that separates the shaded circles from the clear one for f_3 and we conclude that f_3 is a non linearly separable boolean function.

Note that there are four non zero coefficients of $H_3 f_3$ of equal magnitude. Let us choose (arbitrarily) the first three. They correspond to the forms $0,0,1 \leftrightarrow x_3$, $0,1,1 \leftrightarrow x_2 \oplus x_3$, and $1,0,1 \leftrightarrow x_1 \oplus x_3$. Let us now remap f_3 along these new axes as shown in Figure 11. For example, the point $(x_1, x_2, x_3) = (1,1,1)$ maps to $(x_3, x_2 \oplus x_3, x_1 \oplus x_3) = (1,0,0)$ in the new coordinate frame.

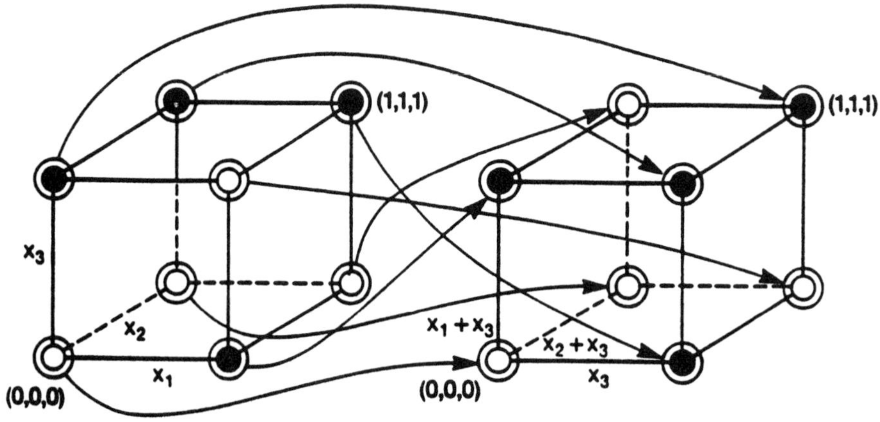

Figure 11 Remapping of f_3

Having done the remapping we observe that f_3 is separable in this new coordinate system. (The reader is encouraged to show that, for this particular function, separability can be achieved by mapping f_3 into any of the four coordinate systems whose bases are 3 out of the 4 forms corresponding to the largest Hadamard transform coefficients.

This experiment suggests a promising method to increase the separating efficiency of a TLU and we propose the following algorithm:

> Given 2^d d-dimensional pattern points, compute the Hadamard transform. Identify the m coefficients of largest magnitude and map the pattern points into the new coordinate system whose basis is defined by the forms corresponding to the identified coefficients. The points will still, in general, not be separable but a TLU operating in this new coordinate system may be made to exhibit a vastly improved separation efficiency.

This *spectrally preconditioned threshold logic unit* is shown in Figure 12. The $\{L_i\}$ are the m modulo-two (linear) sums indicated by the m largest Hadamard transform coefficients.

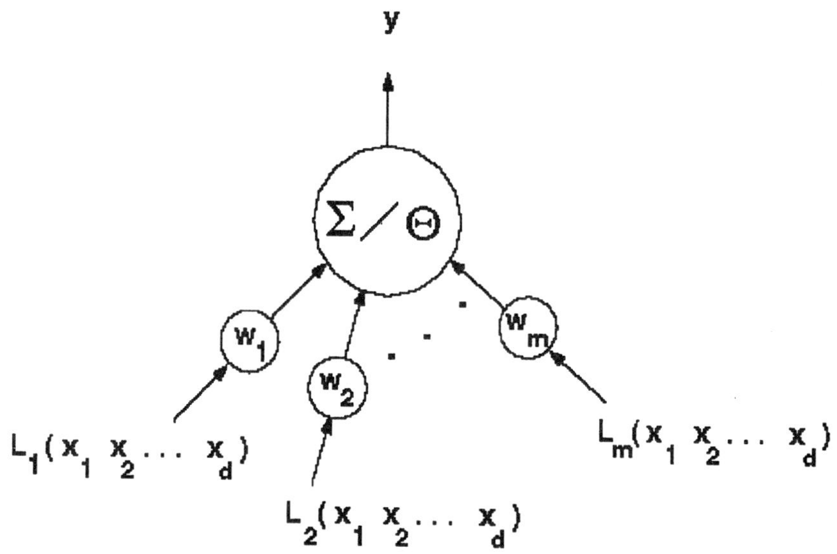

Figure 12 The Spectrally Preconditioned Threshold Logic Unit

Empirical data seems to support the concept. All boolean functions of 2, 3, and 4 variables were examined to see how large m, the number of axes in the new coordinate system needed to be to achieve linear separability. The answer is surprising and exciting and is shown in Table 12.

Table 12 Percentage of all Boolean Functions of Degree 2, 3 and 4 Separable by m or Fewer Axes

m	2 Variables	3 Variables	4 Variables
1	50.00	6.25	0.05
2	100.00	28.13	0.48
3		87.50	3.96
4		100.00	22.46
5			77.34
6			86.13
7			98.44
8			98.44
9			98.63
10			99.41
11			100.00

Next, Monte-Carlo generation of random boolean functions of 5 and 6 variables was undertaken. The same experiments were tried with the condition that the maximum number of axes allowed [max(m)] was set at 11 and the training time limited. The results are shown in Table 13. The reader is urged to recall that only $8.3 \cdot 10^{-11}$ percent of the boolean functions of six variables are separable with a TLU vice approximately 1.2 percent with an eleven axis spectrally preconditioned TLU.

Table 13 Percentage of all Boolean Functions of Degree 5 and 6 Separable by m or Fewer Axes

m	5 Variables	6 Variables
1	0	0
2	0	0
3	0	0
4	0	0
5	1.38	0
6	9.04	0
7	26.92	0
8	43.38	0
9	56.96	0
10	66.88	0.2
11	82.04	1.2

Empirically, the patterns that require a large m for separation through remapping are those functions that have a Hadamard transform that is flat or near flat, ie, equal or approximately equal magnitudes of the coefficients. Rothaus (1976) terms a boolean function *bent* if its Hadamard transform coefficients are all of equal magnitude. Thus $f_2(x_1, x_2) = x_1 x_2$ and $f_4(x_1, x_2, x_3, x_4) = x_1 x_2 + x_3 x_4$ are examples of bent functions. We are therefore interested in the realization of bent or near bent functions via spectrally preconditioned threshold logic.

Drawing from Yarlagadda and Hershey (1989), we begin with a consideration of two arbitrary boolean sequence vectors,[7] s_1 and s_2 each of length 2^n bits. Karpovsky (1976) defines the correlation of these two sequences as

$$\begin{aligned} R_{s_1 s_2}(i) &= \sum_{\tau=0}^{2^n-1} s_1(\tau) s_2(i \oplus \tau) \\ &= s_1(i) \circ s_2(i) \end{aligned} \tag{153}$$

where $s_i(\tau)$ is the τ-th entry in s_i, $i \oplus \tau$ represents bit-by-bit modulo-two addition of the arguments i and τ expressed in normal binary, and $'\circ'$ specifies boolean correlation.

As an example of what we have so far, consider the case for $n = 2$. Table 14 constructs $i \oplus \tau$ for this value of n. In this table, the variables i and τ are expressed in base-10 numbers and also in normal binary (in parentheses).

[7] We will develop the following about the notion of a *sequence*. Later on, we will be considering that sequence to be a truth table, read in a standard way, columnarly, from top to bottom, which defines a boolean function.

Table 14 Computation of $i \oplus \tau$ for n=2

		(τ) 0 (0,0)	1 (0,1)	2 (1,0)	3 (1,1)
(i)	0 (0,0)	0 (0,0)	1 (0,1)	2 (1,0)	3 (1,1)
	1 (0,1)	1 (0,1)	0 (0,0)	3 (1,1)	2 (1,0)
	2 (1,0)	2 (1,0)	3 (1,1)	0 (0,0)	1 (0,1)
	3 (1,1)	3 (1,1)	2 (1,0)	1 (0,1)	0 (0,0)

We can now write the boolean correlation of the two sequences in matrix form as

$$R_{s_1 s_2} = \begin{pmatrix} s_2(0) & s_2(1) & s_2(2) & s_2(3) \\ s_2(1) & s_2(0) & s_2(3) & s_2(2) \\ s_2(2) & s_2(3) & s_2(0) & s_2(1) \\ s_2(3) & s_2(2) & s_2(1) & s_2(0) \end{pmatrix} \begin{pmatrix} s_1(0) \\ s_1(1) \\ s_1(2) \\ s_1(3) \end{pmatrix} \tag{154}$$

or in symbolic form

$$R_{s_1 s_2} = A_{s_2} s_1 \, . \tag{155}$$

Using (153) we can also write

$$\begin{aligned} R_{s_2 s_1}(i) &= s_2(i) \circ s_1(i) \\ &= R_{s_1 s_2}(i) \, . \end{aligned} \tag{156}$$

With this, (153) may be written as

$$R_{s_1 s_2} = A_{s_1} s_2 \tag{157}$$

where A_{s_1} can be obtained from (154) by replacing $s_2(i)$ with $s_1(i)$. The matrices $\{A_{s_i}\}$ have a special structure that can be seen from (154). For illustrative purposes let the matrix B_2 represent the indices of A_{s_2} in (154), ie, $B_2(i,j)$ is the index of the (i,j)-th entry in A_{s_2}. (This is what we computed in Table 14.) So,

$$B_2 = \begin{pmatrix} 0 & 1 & 2 & 3 \\ 1 & 0 & 3 & 2 \\ 2 & 3 & 0 & 1 \\ 3 & 2 & 1 & 0 \end{pmatrix} . \tag{158}$$

We can generalize this and show that B_n may be obtained from B_{n-1} by the recursion

$$B_n = \begin{pmatrix} B_{n-1} & \bar{B}_{n-1} \\ \bar{B}_{n-1} & B_{n-1} \end{pmatrix} \tag{159}$$

with $B_0 = 0$ and $\bar{B}_{n-1}$ is a $2^{n-1} \times 2^{n-1}$ matrix with

$$\bar{B}_{n-1}(i,j) = B_{n-1}(i,j) + 2^{n-1} \, . \tag{160}$$

From (154) we see that the matrix A_{s_2} has the general form

$$A_{s_2}(n) = \begin{pmatrix} A_{s_2}(n-1) & \bar{A}_{s_2}(n-1) \\ \bar{A}_{s_2}(n-1) & A_{s_2}(n-1) \end{pmatrix} \tag{161}$$

where the indices of the variable s_2 are obtained from (159). Equations (160) and (161) follow by construction.

Next, let us consider the eigenvalue-eigenvector decomposition of (161) which is similar to the eigenvalue-eigenvector decomposition of circulant matrices as discussed in Hershey and Yarlagadda (1986). We can state that

$$2^{-n} H_n A_{s_2}(n) H_n = D \tag{162}$$

where

$$D = dia(d_0, d_1, ..., d_{2^n-1}) \tag{163}$$

and $H_n s_2 = col(d_0, d_1, ..., d_{2^n-1})$ .. This can be easily seen using the fact that

$$H_n = \begin{pmatrix} H_{n-1} & 0 \\ 0 & H_{n-1} \end{pmatrix} \begin{pmatrix} I & I \\ I & -I \end{pmatrix} \tag{164}$$

and

$$A_{s_2}(n) = \frac{1}{2} \begin{pmatrix} I & I \\ I & -I \end{pmatrix} \begin{pmatrix} A_{s_2}(n-1) + \bar{A}_{s_2}(n-1) & 0 \\ 0 & A_{s_2}(n-1) - \bar{A}_{s_2}(n-1) \end{pmatrix} \cdot \begin{pmatrix} I & I \\ I & -I \end{pmatrix} \tag{165}$$

where I is an identity matrix of order 2^{n-1}. Continuing with the above process, we can see that the diagonal entries $\{d_i\}$ in (163) are the transform coefficients, and the assertion is demonstrated.

A special case obtains when $s_1 = s_2$ and $R_{s_1 s_2} = A_{s_1} s_1$ which produces the autocorrelation coefficients. Using (162) we have$R_{s_1 s_1} = 2^{-n} H_n D H_n s_1$ and $H_n R_{s_1 s_1} = D H_n s_1$ which shows that the Hadamard transform of the autocorrelation coefficients gives the magnitude spectrum of the sequence vector s_i. Similar to the discrete Fourier transform, the auto (cross) correlation coefficients can be obtained using transforms. The procedure is:

i. Find the Hadamard transform of s_1: $F = H_n s_1$,
ii. Compute the square of the magnitude spectrum,
iii. Compute the inverse transform.

As an example, consider $s = 1, 1, 1, -1$. Table 15 displays the results.

Table 15 Computing the magnitude and Autocorrelation of a Sequence

s	s_0	$H_2 s_0$	Magnitude Spectrum	$R_{s_0 s_0}$
1	0	1	1	1
1	0	-1	1	0
1	0	-1	1	0
-1	1	1	1	0

The column denoted s_0 is obtained from s by simply changing $1's$ to $0's$ and $-1's$ to $1's$.

So far this chapter has been concerned with boolean functions. The results can clearly also be considered in the real domain. In (162), the autocorrelation matrix is diagonalized by using the Hadamard matrix. In the real domain the interest may be in the autocorrelation matrix and its diagonalization. Clearly the diagonalization of an arbitrary matrix is a major task. There are very few matrices that allow a simple eigenvalue-eigenvector decomposition. For example, circulant matrices have the Discrete Fourier Transform (DFT) matrix as a modal matrix. The eigenvalues can be computed by taking the DFT of the first column or the first row of the circulant matrix. [See p. 189 of Hershey and Yarlagadda (1986).] A natural question arises and that concerns what can be said about matrices that have the Hadamard matrix as a modal matrix.

To investigate, let us consider the product $\overline{A_n} = H_n \Lambda^{(n)} H_n$ and $\frac{1}{2^n}\overline{A_n}$. The matrix $\overline{A_n}$ has a structure. It has at most 2^n distinct entries. For example, the diagonal entries are all equal. For others, we will prove that

$$H_n \Lambda^{(n)} H_n = \sum_{i=0}^{2^n-1} D_i^{(n)} x_i^{(n)} \tag{166}$$

where the $D_i^{(n)}$'s are constants and the x_i's are constructed from 2×2 matrices

$$\begin{aligned} x_0^{(1)} &= \begin{pmatrix} 1 & 0 \\ 0 & 1 \end{pmatrix} \\ x_1^{(1)} &= \begin{pmatrix} 0 & 1 \\ 1 & 0 \end{pmatrix}, \end{aligned} \tag{167}$$

$$
\begin{aligned}
x_0^{(2)} &= x_0^{(1)} \otimes x_0^{(1)} = \begin{pmatrix} x_0^{(1)} & 0 \\ 0 & x_0^{(1)} \end{pmatrix}, \\
x_1^{(2)} &= x_0^{(1)} \otimes x_1^{(1)} = \begin{pmatrix} x_1^{(1)} & 0 \\ 0 & x_1^{(1)} \end{pmatrix}, \\
x_2^{(2)} &= x_1^{(1)} \otimes x_0^{(1)} = \begin{pmatrix} 0 & x_0^{(1)} \\ x_0^{(1)} & 0 \end{pmatrix}, \\
x_3^{(2)} &= x_1^{(1)} \otimes x_1^{(1)} = \begin{pmatrix} 0 & x_1^{(1)} \\ x_1^{(1)} & 0 \end{pmatrix}.
\end{aligned}
\tag{168}
$$

At the k-th stage we have $\left\{x_0^{(k)}, \ldots, x_{2^k-1}^{(k)}\right\}$ and at the $(k+1)$-st stage $\left\{x_0^{(k+1)}, \ldots, x_{2^k-1}^{(k+1)}, x_{2^k}^{(k+1)}, \ldots, x_{2^{k+1}-1}^{(k+1)}\right\}$ where

$$
x_i^{(k+1)} = \begin{cases} \begin{pmatrix} x_i^{(k)} & 0 \\ 0 & x_i^{(k)} \end{pmatrix} = x_0^{(1)} \otimes x_i^{(k)}, \; i = 0, 1, \ldots, 2^k - 1 \\ \begin{pmatrix} 0 & x_i^{(k)} \\ x_i^{(k)} & 0 \end{pmatrix} = x_1^{(1)} \otimes x_i^{(k)}, \; i = 2^k, 2^k + 1, \ldots, 2^{k+1} - 1 . \end{cases}
\tag{169}
$$

Clearly,

$$
\sum_{i=0}^{2^n-1} x_i^{(n)} = H_n .
\tag{170}
$$

We can now prove

$$
\begin{aligned}
H_n \Lambda^{(n)} H_n &= \overline{A_n} \\
&= \sum_{i=0}^{2^n-1} D_i^{(n)} x_i^{(n)}
\end{aligned}
\tag{171}
$$

by induction, where the $\left\{D_i^{(n)}\right\}$ are constants are the $\left\{x_i^{(n)}\right\}$ are matrices. The constants $\left\{D_i^{(n)}\right\}$ are the entries in the first column of $\overline{A_n}$ and the matrices $\left\{x_i^{(n)}\right\}$ are defined in (169).

First,

$$
\begin{aligned}
H_1 \Lambda^{(1)} H_1 &= \begin{pmatrix} +1 & +1 \\ +1 & -1 \end{pmatrix} \begin{pmatrix} \lambda_1^{(1)} & 0 \\ 0 & \lambda_2^{(1)} \end{pmatrix} \begin{pmatrix} +1 & +1 \\ +1 & -1 \end{pmatrix} \\
&= \begin{pmatrix} D_0^{(1)} & D_1^{(1)} \\ D_1^{(1)} & D_0^{(1)} \end{pmatrix} \\
&= \begin{pmatrix} \lambda_1^{(1)} + \lambda_2^{(1)} & \lambda_1^{(1)} - \lambda_2^{(1)} \\ \lambda_1^{(1)} - \lambda_2^{(1)} & \lambda_1^{(1)} + \lambda_2^{(1)} \end{pmatrix} \\
&= \left(\lambda_1^{(1)} + \lambda_2^{(1)}\right) \begin{pmatrix} 1 & 0 \\ 0 & 1 \end{pmatrix} + \left(\lambda_1^{(1)} - \lambda_2^{(1)}\right) \begin{pmatrix} 0 & 1 \\ 1 & 0 \end{pmatrix} \\
&= D_0^{(1)} x_0^{(1)} + D_1^{(1)} x_1^{(1)}
\end{aligned}
\tag{172}
$$

indicating that the result is true *for* $n = 1$.

We assume that the result is true for $n = k-1$, ie,

$$H_{k-1}\Lambda^{(k-1)}H_{k-1} = \sum_{i=0}^{2^k-1} D_i^{(k-1)} x_i^{(k-1)} \tag{173}$$

and we will prove that the result is true for $n = k$:

$$\begin{aligned}
H_k\Lambda^{(k)}H_k &= \begin{pmatrix} +H_{k-1} & +H_{k-1} \\ +H_{k-1} & -H_{k-1} \end{pmatrix} \begin{pmatrix} \Lambda_1^{(k)} & 0 \\ 0 & \Lambda_2^{(k)} \end{pmatrix} \begin{pmatrix} +H_{k-1} & +H_{k-1} \\ +H_{k-1} & -H_{k-1} \end{pmatrix} \\
&= \begin{pmatrix} H_{k-1}\left(\Lambda_1^{(k)} + \Lambda_2^{(k)}\right)H_{k-1} & H_{k-1}\left(\Lambda_1^{(k)} - \Lambda_2^{(k)}\right)H_{k-1} \\ H_{k-1}\left(\Lambda_1^{(k)} - \Lambda_2^{(k)}\right)H_{k-1} & H_{k-1}\left(\Lambda_1^{(k)} + \Lambda_2^{(k)}\right)H_{k-1} \end{pmatrix} \\
&= \begin{pmatrix} \sum_{i=0}^{2^{k-1}-1} D_{1i}^{(k-1)} x_i^{(k-1)} & \sum_{i=0}^{2^{k-1}-1} D_{2i}^{(k-1)} x_i^{(k-1)} \\ \sum_{i=0}^{2^{k-1}-1} D_{2i}^{(k-1)} x_i^{(k-1)} & \sum_{i=0}^{2^{k-1}-1} D_{1i}^{(k-1)} x_i^{(k-1)} \end{pmatrix} \\
&= \begin{pmatrix} \sum_{i=0}^{2^{k-1}-1} D_{1i}^{(k-1)} x_i^{(k-1)} & 0 \\ 0 & \sum_{i=0}^{2^{k-1}-1} D_{1i}^{(k-1)} x_i^{(k-1)} \end{pmatrix} \\
&\quad + \begin{pmatrix} 0 & \sum_{i=0}^{2^{k-1}-1} D_{2i}^{(k-1)} x_i^{(k-1)} \\ \sum_{i=0}^{2^{k-1}-1} D_{2i}^{(k-1)} x_i^{(k-1)} & 0 \end{pmatrix} \\
&= \sum_{i=0}^{2^{k-1}-1} D_{1i}^{(k-1)} \begin{pmatrix} x_i^{(k-1)} & 0 \\ 0 & x_i^{(k-1)} \end{pmatrix} \\
&\quad + \sum_{i=0}^{2^{k-1}-1} D_{2i}^{(k-1)} \begin{pmatrix} 0 & x_i^{(k-1)} \\ x_i^{(k-1)} & 0 \end{pmatrix} \\
&= \sum_{i=0}^{2^k-1} D_i^{(k)} x_i^{(k)}
\end{aligned} \tag{174}$$

where

$$D_i^{(k)} = \begin{cases} D_{1i}^{(k-1)}, & i=0,1,2,\ldots,2^{k-1}-1 \\ D_{1i}^{(k-1)}, & i=2^{k-1},2^{k-1}+1,2^{k-1}+2,\ldots,2^k-1 \end{cases} \tag{175}$$

thus proving the assertion that

$$\begin{aligned} A &= H_n\Lambda^{(n)}H_n \\ &= \sum_{i=0}^{2^n-1} D_i^{(n)} x_i^{(n)} . \end{aligned} \tag{176}$$

As an example, pick $N = 8$, $n = 3$, and let $D_i^{(3)} = D_i$ in the following:

$$A = \frac{1}{N}\begin{pmatrix} D_0 & D_1 & D_2 & D_3 & D_4 & D_5 & D_6 & D_7 \\ D_1 & D_0 & D_3 & D_2 & D_5 & D_4 & D_7 & D_6 \\ D_2 & D_3 & D_0 & D_1 & D_6 & D_7 & D_4 & D_5 \\ D_3 & D_2 & D_1 & D_0 & D_7 & D_6 & D_5 & D_4 \\ D_4 & D_5 & D_6 & D_7 & D_0 & D_1 & D_2 & D_3 \\ D_5 & D_4 & D_7 & D_6 & D_1 & D_0 & D_3 & D_2 \\ D_6 & D_7 & D_4 & D_5 & D_2 & D_3 & D_0 & D_1 \\ D_7 & D_6 & D_5 & D_4 & D_3 & D_2 & D_1 & D_0 \end{pmatrix} . \tag{177}$$

From this, we can see that there are only 8 distinct elements. It is interesting to note that the subscripts in (177) follow the pattern in (159) — (160).

The eigenvalues of the matrix A can be easily computed from the first column of the matrix A. First,

$$\begin{aligned} A &= \frac{1}{N} H_n \Lambda H_n \\ &= \frac{1}{N} \overline{A} \end{aligned} \tag{178}$$

or

$$\begin{aligned} H_n A &= \Lambda H_n \\ &= \begin{pmatrix} \lambda_0 & & & \\ & \lambda_1 & & \\ & & \ddots & \\ & & & \lambda_{2^n-1} \end{pmatrix} H_n . \end{aligned} \tag{179}$$

Let

$$A = (\, \underline{a}_0 \quad \underline{a}_1 \quad \cdots \quad \underline{a}_{2^n-1} \,) \tag{180}$$

where $\underline{a}_i$ is the i-th column in A. Then

$$\begin{aligned} H_n \underline{a}_0 &= \Lambda \begin{pmatrix} 1 \\ 1 \\ \vdots \\ 1 \end{pmatrix} \\ &= \begin{pmatrix} \lambda_0 \\ \lambda_1 \\ \vdots \\ \lambda_{2^n-1} \end{pmatrix} \end{aligned} \tag{181}$$

indicating that the eigenvalues of A can be obtained by taking the Hadamard transform of the first column in A. The matrix A is highly structured. Relating this matrix to an arbitrary correlation matrix is obviously signal dependant. Noting this and that the entries of H_n are ± 1, the effectiveness of a Hadamard transform as a tool for compression is limited compared to the DFT. However, due to the simplicity of implementing the Hadamard transform, it is an attractive tool for quick analysis.

Chapter 16 Synthesis of Bent Functions

The empirical results of the last section indicate that the question of synthesis of bent functions (Rothaus, 1976) may be worthwhile since they seem to require the greatest number of dimensions, m, in our approach of spectrally preconditioned threshold logic.

As a rather minor first point, but one which will make the discussion somewhat easier, let us consider the consequence of boolean function representation in terms of zeros and ones vice ones and minus ones. Using the notation of the example at the end of the last section, we can express s, which can be considered as the output of an *'AND'* function, a bent boolean function, as $s = (1_c - 2s_0)$ or, equivalently, $s_0 = \frac{1}{2}(1_c - s)$ where 1_c is a column vector of all ones of dimension 2^n, s is a column vector of ones and minus ones, and s_0 exhibits only ones and zeros. Now

$$H_n s_0 = \frac{1}{2}(H_n 1_c - H_n s) \; . \tag{182}$$

Noting that $H_n 1_c = col(2^n, 0, ..., 0)$ and assuming that s is a bent sequence vector with more positive ones than negative ones, we have

$$(H_n s_o)_i = \begin{cases} \frac{1}{2}(2^n - 2^{n/2}), \; i=0 \\ \mp 2^{\left(\frac{n}{2}\right)-1}, \; i \neq 0 \end{cases} \tag{183}$$

and the autocorrelation coefficients[8] are given by

$$R(i) = \begin{cases} b + 2^{n-2}, \; i=0 \\ b, \; 1 \leq i \leq 2^n - 1 \end{cases} \tag{184}$$

where $b = \frac{1}{4}\left[2^n - 2(2^{n/2})\right]$. We will use these results when we consider the bounds on the number of weights required for threshold realization of bent functions.[9]

The basic problem is to find a set of weights $\{w_i\}$, $i = 0, 1, 2, ..., 2^n - 1$ such that $H_n w = s$ where $w^T = (\, w_0 \quad w_1 \quad w_2 \quad \dots \quad w_{2^n - 1} \,)$ is the weight vector and $s^T = (\, s_0 \quad s_1 \quad s_2 \quad \dots \quad s_{2^n-1} \,)$ is the Boolean vector that is taken from the truth table and so $\{s_i\} \in \{\pm 1\}$. Noting that the inverse of H_n is $2^{-n} H_n$ we see that the weights are simply proportional to the transform coefficients of s. The solution is, of course, unique, but what we are really after is slightly different.

Given

$$H_n w = p \tag{185}$$

with

$$p^T = (\, p_0 \quad p_1 \quad p_2 \quad \dots \quad p_{2^n - 1} \,) \tag{186}$$

---

8 The subscripts on R are omitted for simplicity.

9 The case wherein s has more minus ones than ones is a trivial extension.

such that the hard quantized values of the $\{p_i\}$, $Q_u\{p_i\}$, where

$$Q_u(p_i) = \{{}^{1\ ,\ p_i>0}_{-1\ ,\ p_i<0} \quad i = 0, 1, 2, ..., 2^n - 1 \tag{187}$$

or

$$Q_u(p_i) = s_i \ . \tag{188}$$

The problem is to find the 'best' set of weights in some sense such that (185) — (187) are satisfied. The best set may be in the form that the number of non zero weights is a minimum, or that the weights have minimum integer values, and so on.

The problem of finding the smallest set of non zero weights is of most interest and so far as we know remains an open question (Picton, 1981). Normally, the first weight, w_0 is usually not included in the weight count. For certain special cases, Dertouzos (1965) and Chow (1961) have characterized threshold functions using spectral coefficients. Others, such as Golomb (1959) classified Boolean functions and synthesized functions based on these classifications. (See Edwards (1974)). The general realization in terms of minimum number of weights for bent functions is, as far as we know, not known. However, for some special cases, this has been discussed by Edwards (1974). Olsen et al (1982) presented families of bent function sequences that can be generated through non linear operations on m-sequences.

It will be more convenient and traditional for us to state the problem in terms of zeros and ones. Conventionally, in the pattern recognition area, zeros and ones are used in the coefficient matrix (185) while still keeping the sequence vector s in terms of ones and minus ones. To derive this formulation we write (185) as

$$(\, 1_c \quad \acute{H}_n \,) \begin{pmatrix} w_0 \\ \acute{w} \end{pmatrix} = p \tag{189}$$

where 1_c is a column vector of dimension 2^n of all ones. $\acute{H}_n$ is obtained from H_n by deleting the first column, and $\acute{w}$ is obtained from w by deleting the first entry. Let U be a $2^n \times (2^n - 1)$ matrix with all ones and let (189) be rewritten as

$$-\frac{1}{2} U \acute{w} + \frac{1}{2} U \acute{w} - \acute{H}_n \left(-\frac{1}{2} \acute{w} \right) + 1_c \frac{1}{2} w_0 = \frac{1}{2} p \tag{190}$$

where we have added and subtracted the term $U\acute{w}$ and divided each term in the equation by 2. We may simplify (190) to

$$\left(\frac{1}{2} w_0 + \frac{1}{2} \sum_{i \neq 0} w_i \right) 1_c - \frac{1}{2} \left(\acute{H}_n - U \right) (-\acute{w}) = \frac{1}{2} p \tag{191}$$

or

$$\left(1_c - \frac{1}{2} \left(\acute{H}_n - U \right) \right) \begin{pmatrix} \frac{1}{2} \sum_i w_i \\ -\acute{w} \end{pmatrix} = \frac{1}{2} p \tag{192}$$

which now has the coefficient matrix with zeros and ones. For various other types of transformations see Hurst (1973).

As an example, consider the sequence $s^T = (\,1 \quad 1 \quad 1 \quad -1\,)$. Equation (185) can be expressed as

$$\begin{pmatrix} +1 & +1 & +1 & +1 \\ +1 & -1 & +1 & -1 \\ +1 & +1 & -1 & -1 \\ +1 & -1 & -1 & +1 \end{pmatrix} \begin{pmatrix} w_0 \\ w_1 \\ w_2 \\ w_3 \end{pmatrix} = \begin{pmatrix} p_0 \\ p_1 \\ p_2 \\ p_3 \end{pmatrix} \tag{193}$$

where $Q_u(p_i) = s_i$. Now, (192) can be expressed by

$$\begin{pmatrix} 1 & 0 & 0 & 0 \\ 1 & 1 & 0 & 1 \\ 1 & 0 & 1 & 1 \\ 1 & 1 & 1 & 0 \end{pmatrix} \begin{pmatrix} w_0' \\ w_1' \\ w_2' \\ w_3' \end{pmatrix} = \frac{1}{2} \begin{pmatrix} p_0 \\ p_1 \\ p_2 \\ p_3 \end{pmatrix} \tag{194}$$

where

$$w_0' = \frac{1}{2} \sum_{i=0}^{3} w_i \tag{195}$$

and

$$w_i' = -w_i \,, \quad i \neq 0 \,. \tag{196}$$

Because equations (185) and (192) are directly related, we need to discuss the synthesis problem for only one case. Let us now consider some bounds for m, the minimum number of non zero weights required to realize a Boolean function.

We know of no explicit formula for finding m, however, Karpovsky (1976) provides the following upper bound: *For any completely specified Boolean function of n arguments,*

$$m \leq 2^n - max\left[2^{n-1} - R(0) + \underset{i \neq 0}{maxR(i)}, R(0) - \underset{i \neq 0}{minR(i)}\right] \tag{197}$$

where $R(i)$ is the autocorrelation function discussed earlier.

For bent functions, (197) reduces via (184) to

$$\begin{aligned} m &\leq 2^n - max\left[2^{n-1} - \left(b + 2^{n-2}\right) + b, b + 2^{n-2} - b\right] \\ &\leq 2^n - 2^{n-2} \\ &= 3\left(2^{n-2}\right) . \end{aligned} \tag{198}$$

By definition, the bent sequences have a constant magnitude spectrum. The magnitude of the difference between the number of positive and negative entries in the bent sequence and also in its transform is a constant and equal to $2^{n/2}$ for a sequence of length 2^n. This allows a simple realization with the non zero weights all equal to unity.

We let s be a 2^n dimensional bent sequence and c_k be the k-th entry in s, ie, $c_k = (s)_k$, which is either one or minus one. Further, let h_k , $0 \leq k \leq 2^n - 1$, be the column vectors in the Hadamard matrix. Then, the vector

$$b_k = \frac{1}{2}\left(c_k h_k + 2^{-n/2} H_n s\right) \tag{199}$$

contains ones, minus ones, and zeros. Now,

$$\begin{aligned} H_n b_k &= \frac{1}{2} c_k H_n h_k + \frac{1}{2} \frac{1}{2^{n/2}} H_n^2 s \\ &= \frac{1}{2} 2^n c_k (0, ..., 0, 1, 0, ..., 0)^T + \frac{1}{2} 2^{n/2} s \end{aligned} \tag{200}$$

where the single one is in the k-th location in the first term. From (200) it follows that s and $H_n b_k$ have the same sign pattern disclosing that b_k realizes the bent sequence s with

$$N_1 = \frac{2^n + 2^{n/2}}{2} \tag{201}$$

non zero weights. Note that N_1 corresponds to the number of non zero entries in b_k. This follows from the fact that all the entries in b_k are either plus or minus one or zero and

$$\begin{aligned} b_k^T b_k &= \frac{1}{4}\left(c_k^2 h_k^T h_k + \frac{2}{2^{n/2}} c_k h_k^T H_n s + \frac{1}{2^n} s^T H_n^2 s \right) \\ &= \frac{1}{4}\left(2 \cdot 2^n + 2 \cdot 2^{n/2}\right) . \end{aligned} \tag{202}$$

A special case of b_k is b_0 wherein all the entries in b_0 are either all ones, minus ones, or zeros.

As an example, consider the four vectors s_1, s_2, s_3, and s_4 and their realizations displayed in Table 16. We have used $b_k^{(i)}$ to indicates that sequence i $(1 \leq i \leq 4)$ is under consideration and k identifies the particular b in (199).

Table 16 Realizations of Vectors s_1, s_2, s_3, and s_4

s_1	$b_0^{(1)}$	$H_2 b_0^{(1)}$	s_2	$b_2^{(2)}$	$H_2 b_2^{(2)}$	s_3	$b_2^{(3)}$	$H_2 b_2^{(3)}$	s_4	$b_0^{(4)}$	$H_2 b_0^{(4)}$
1	1	3	1	0	1	1	1	1	-1	0	-3
1	1	1	1	-1	1	-1	1	-1	1	-1	1
1	1	1	-1	1	-3	1	-1	3	1	-1	1
-1	0	-1	1	1	1	1	0	1	1	-1	1

There are clearly other realizations that can be used to achieve other constraints, such as the first weight being non zero, and so on. Note that all the realizations in Table 16 have three non zero weights.

The bent sequence realization given in (199) is simple and the weight vector contains only plus and minus ones and zeros. The number of non zero weights by the above procedure is N_1 as given in (201). When the length of the bent sequence vector is larger than four, the sequences can be realized with a fewer number of weights than N_1. This can be seen from the fact that the magnitude of the entries in $H_n b_k$ are greater than or equal to $2^{(n/2)-1}$ by (200). Therefore, by forcing $2^{(n/2)-1} - 1$ non zero entries in b_k to zero, we will have a realization with

$$N_2 = N_1 - 2^{(n/2)-1} + 1 \tag{203}$$

number of non zero weights.

As another example, consider the bent sequence vector $s = (1 \quad 1 \quad 1 \quad 1 \quad 1 \quad -1 \quad 1 \quad -1 \quad 1 \quad 1 \quad -1 \quad -1 \quad 1 \quad -1 \quad -1 \quad 1)^T$ and its realization. Table 17 displays s, b_0, $H_4 b_0$, $b_0(16)$, and $H_4 b_0(16)$ where $b_0(16)$ is formed from b_0 by setting the sixteenth element in b_0 to zero.

Table 17 Realization of Bent Sequence Vector s

s^T	1	1	1	1	1	-1	1	-1	1	1	-1	-1	1	-1	-1	1
b_0^T	1	1	1	1	1	0	1	0	1	1	0	0	1	0	0	1
$(H_4 b_0)^T$	10	2	2	2	2	-2	2	-2	2	2	-2	-2	2	-2	-2	2
$b_0^T(16)$	1	1	1	1	1	0	1	0	1	1	0	0	1	0	0	0
$(H_4 b_0(16))^T$	9	3	3	1	3	-3	1	-1	3	1	-3	-1	1	-1	-1	1

From Table 17 we see that $H_4 b_0(16)$ matches the sign pattern of s and therefore $b_0(16)$ realizes s. The vector s that we have used was created by simply stacking the columns of H_4. We can realize such vectors by using the recursive structure:

$$\begin{aligned} &\begin{pmatrix} H_{(n/2)-1} & H_{(n/2)-1} \\ H_{(n/2)-1} & -H_{(n/2)-1} \end{pmatrix} \begin{pmatrix} A_{(n/2)-1} & A_{(n/2)-1} \\ A_{(n/2)-1} & 0 \end{pmatrix} \\ &\cdot \begin{pmatrix} H_{(n/2)-1} & H_{(n/2)-1} \\ H_{(n/2)-1} & -H_{(n/2)-1} \end{pmatrix} \\ = &\begin{pmatrix} 3B_{(n/2)-1} & B_{(n/2)-1} \\ B_{(n/2)-1} & -B_{(n/2)-1} \end{pmatrix} \\ = &B_n \end{aligned} \tag{204}$$

where

$$B_{(n/2)-1} = H_{(n/2)-1} A_{(n/2)-1} H_{(n/2)-1} \tag{205}$$

and $A_{(n/2)-1}$ realizes $H_{(n/2)-1}$. With

$$A_2 = \begin{pmatrix} 1 & 1 \\ 1 & 0 \end{pmatrix} \tag{206}$$

we have

$$A_4 = \begin{pmatrix} 1 & 1 & 1 & 1 \\ 1 & 0 & 1 & 0 \\ 1 & 1 & 0 & 0 \\ 1 & 0 & 0 & 0 \end{pmatrix} \tag{207}$$

and

$$B_4 = \begin{pmatrix} 9 & 3 & 3 & 1 \\ 3 & -3 & 1 & -1 \\ 3 & 1 & -3 & -1 \\ 1 & -1 & -1 & 1 \end{pmatrix} . \tag{208}$$

From (207)-(208) we see that $b_0(16)$ and $H_4 b_0(16)$ can be obtained. From (204)—(205) we can see that sequences generated from Hadamard matrices can be synthesized with $3^{n/2}$ weights which coincide with N^2 when $n = 4$ and much fewer when n is larger than four. Obviously, there must be other bent sequences that can be realized with $3^{n/2}$ weights. The question is then posed, "What is the lower bound on the minimum number of weights required to realize a given bent sequence?" We *conjecture* that this number is $N_{min} = 3^{n/2}$ corresponding to a 2^n length bent sequence.

Finally, we show that bent sequences which are generated by Kronecker products are realizable using $3^{n/2}$ weights. Recall that in Table 16 we have a realizations for $s_1 = (1 \quad 1 \quad 1 \quad -1)^T$ Let us write this realization in matrix form:

$$\begin{aligned} &\begin{pmatrix} h_0^{(2)} & h_1^{(2)} & h_2^{(2)} \end{pmatrix} \begin{pmatrix} w_0 \\ w_1 \\ w_2 \end{pmatrix} \\ &= \begin{pmatrix} p_0 \\ p_1 \\ p_2 \\ p_3 \end{pmatrix} \\ &= p \end{aligned} \tag{209}$$

where the $\left\{ h_i^{(2)} \right\}$ are columns in H_2, $w_0 = w_1 = w_2 = 1$, and $p_0 = 3$, $p_1 = p_2 = 1$, and $p_3 = -1$ from Table 16. Equation (209) can now be written as

$$\sum_{i=0}^{2} h_i^{(2)} w_i = p . \tag{210}$$

Now consider the realization

$$z = s_1 \otimes s_1 . \tag{211}$$

From (210) we can write

$$
\begin{aligned}
p \otimes p = \sum_{i=0}^{2} h_i^{(2)} w_i \otimes \sum_{i=0}^{2} h_j^{(2)} w_j \\
\sum_{i=0}^{2} \sum_{j=0}^{2} \left(h_i^{(2)} \otimes h_j^{(2)} \right) w_i w_j
\end{aligned}
\tag{212}
$$

or

$$
p \otimes p = R' w \tag{213}
$$

where

$$
R' = \begin{pmatrix} h_0^{(4)} & h_1^{(4)} & h_2^{(4)} & h_4^{(4)} & h_5^{(4)} & h_6^{(4)} & h_8^{(4)} & h_9^{(4)} & h_{10}^{(4)} \end{pmatrix} \tag{214}
$$

and

$$
w^T = \begin{pmatrix} w_0^2 & w_0 w_1 & w_0 w_2 & w_1 w_0 & w_1^2 & w_1 w_2 & w_2 w_0 & w_2 w_1 & w_2^2 \end{pmatrix} . \tag{215}
$$

Note that the $h_i^{(4)}$ in (214) are columns in H_4. In Table 18 we show the result of inserting the zero weights corresponding to $h_i^{(4)}$ that are not in (214), the weight vector, the vector z in (211), and the vector $p \otimes p$.

Table 18 Weight Vector, Vector $p \otimes p$, and Vector z

Weight Vector	1	1	1	0	1	1	1	0	1	1	1	0	0	0	0	0
$(p \otimes p)^T$	9	3	3	-3	1	1	1	-1	3	1	1	-1	-3	-1	-1	1
$(s_1 \otimes s_1)^T$	1	1	1	-1	1	1	1	-1	1	1	1	-1	-1	-1	-1	1

Note that from Table 18, $p \otimes p$ and $s_1 \otimes s_1$ match in sign in the nine locations corresponding to the nine non zero weights. Since we have decomposed a sixteenth order vector into a Kronecker product of two minimal size (four) bent sequence vectors, we conjecture that the sequence z in (211) cannot be realized with fewer than nine weights. In general, if

$$
z = s_1 \otimes s_1 \otimes \cdots \otimes s_1 \tag{216}
$$

corresponding to a 2^n dimensional "basic" bent sequence, we can realize it with $3^{n/2}$ weights and, again, *conjecture* that it cannot be realized with fewer number of weights.

Chapter 17 The m-Sequence

A recursion that has been of exceptional value to the electrical engineering discipline as a whole and to the specific arts of communications, radar and signal processing in specific, is

$$x(m) = x(m-\delta_1)\cdot x(m-\delta_2)\cdots x(m-\delta_r)\cdot x(m-n) \tag{217}$$

where the $\{x(i)\} \in \{\pm 1\}$ and the $\{\delta_i\}$ are integers with $0 < \{\delta_i\} < n$. It is clear that the sequence produced, $\{x(m)\}$, $m = 0, 1, 2, \ldots$ is periodic and can be extended forwards *and* backwards as far as desired. Under certain conditions, the sequence period will be $2^n - 1$.

We associate a polynomial with zero and one coefficients with the recursion (217). This polynomial is

$$f = x^n + x^{\delta_1} + x^{\delta_2} \cdots + x^{\delta_r} + 1 \, . \tag{218}$$

If we were to substitute n consecutive +1's into the recursion (217), then it is clear that the recursion will produce only +1's, forwards or backwards. We then can immediately upperbound the period of a sequence produced by the recursion; this upperbound is $2^n - 1$ steps. The fact that a sequence can be produced for any n, that has a period equal to the upperbound is both remarkable and beautiful. Such a sequence is called a *maximal length sequence* or *m-sequence*. We believe that the latter term was bestowed by Zierler (1959). A comprehensive review of m-sequence architectures and implementations may be found in Chapter 8 of Hershey and Yarlagadda (1986).

A set of $\{\delta_i\}$ for a maximal length sequence for many values of n can be found in Stahnke (1973). The polynomial (218) corresponding to such a set of $\{\delta_i\}$ is termed a *primitive* polynomial. There are $\frac{\phi(2^n-1)}{n}$ primitive polynomials of degree n.[10]

There are two m-sequences of period seven. They are produced by the following two polynomials of degree $n = 3$: $f_1 = x^3 + x^2 + 1$ and $f_2 = x^3 + x + 1$. The seven possible *phases* of the m-sequences for these polynomials are displayed in Table 19.

[10] The function $\phi(\cdot)$ is the Euler totient function. It is a *weakly multiplicative* function in that $\phi(ab) = \phi(a)\phi(b)$ if a and b are relatively prime, ie, $(a, b) = 1$. We may calculate the function for any natural number using the weakly multiplicative property and the definition $\phi(1) = 1$ and $\phi(p^n) = p^{n-1}(p-1)$ for p, a prime.

Table 19 The Seven Phases of the Two m-Sequences of Period 7

$x^3 + x^2 + 1$							$x^3 + x + 1$						
+1	+1	-1	+1	-1	-1	-1	+1	+1	-1	-1	-1	+1	-1
+1	-1	+1	-1	-1	-1	+1	+1	-1	-1	-1	+1	-1	+1
-1	+1	-1	-1	-1	+1	+1	-1	-1	-1	+1	-1	+1	+1
+1	-1	-1	-1	+1	+1	-1	-1	-1	+1	-1	+1	+1	-1
-1	-1	-1	+1	+1	-1	+1	-1	+1	-1	+1	+1	-1	-1
-1	-1	+1	+1	-1	+1	-1	+1	-1	+1	+1	-1	-1	-1
-1	+1	+1	-1	+1	-1	-1	-1	+1	+1	-1	-1	-1	+1

Notice that none of the phases corresponding to one primitive polynomial are contained in the list for the other primitive polynomial.

An m-sequence has a number of very interesting and useful properties. For one, it is pseudorandom in that its cyclic autocorrelation is bi-valued as displayed in Figure 13.

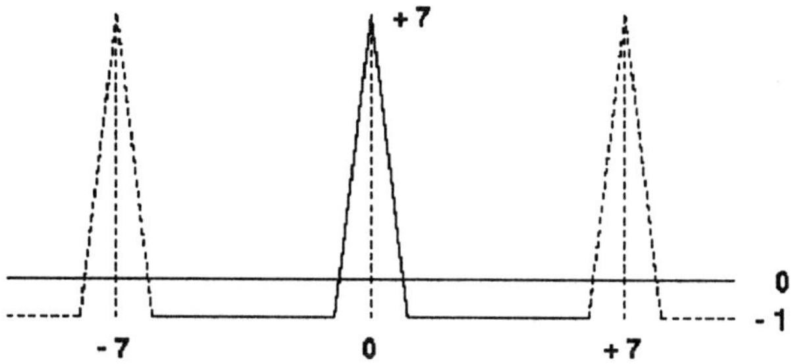

Figure 13 Cyclic Autocorrelation of Period 7-Long m-Sequence

An m-sequence also has the property that if decimated by two, and, consequently, by a power of two, the resulting sequence will be another phase of the same decimated m-sequence. See the example in Figure 14. (For further discussion, see Weathers (1972), pp. 12–13, for example.)

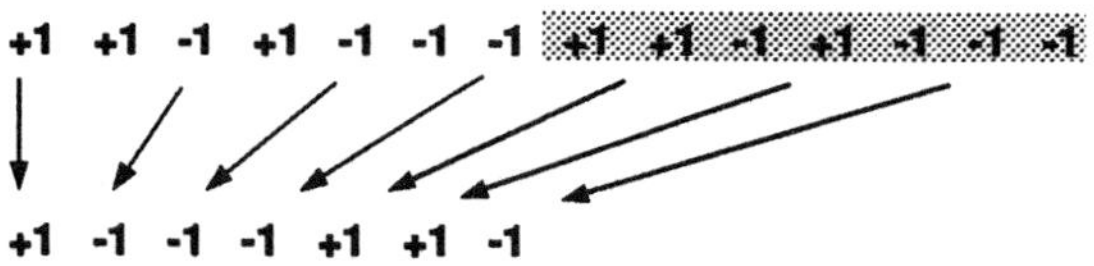

Figure 14 Demonstration of Decimation by Two of the Sequence Generated by $x^3 + x^2 + 1$

We see indeed that the decimation has resulted in another phase of the m-sequence, ie, $+1 \quad -1 \quad -1 \quad -1 \quad +1 \quad +1 \quad -1$ is found in the list of the phases for x^3+x^2+1 and *not* in the list for $x^3 + x + 1$.

There is a remarkable property of m-sequences disclosed by Gold (1966). This property is the existence of a *characteristic sequence* or phase of an m-sequence such that if decimated by two leaves the sequence invariant. The characteristic sequence for x^3+x^2+1 is $-1 \quad +1 \quad +1 \quad -1 \quad +1 \quad -1 \quad -1$ and its invariance to decimation by two is demonstrated in Figure 15.

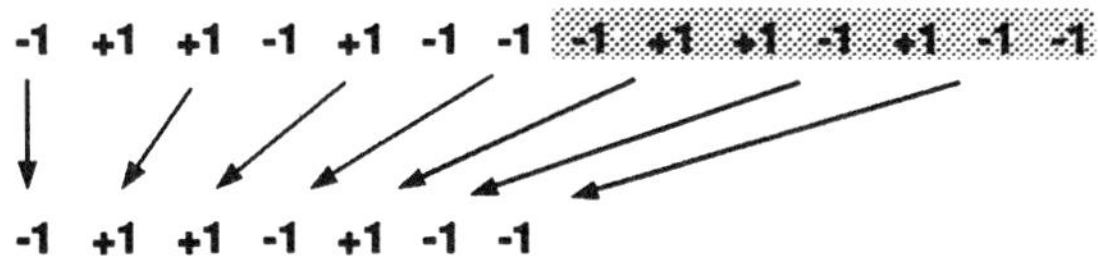

Figure 15 Demonstration of the Characteristic Sequence Corresponding to x^3+x^2+1

Gold (1966) showed that the characteristic sequence could be found for the primitive polynomial f by performing the division

$$\begin{aligned} &\frac{1}{f}\frac{d(xf)}{x}\,, \; n \; odd \\ &\frac{1}{f}\frac{d(xf)}{x}+1\,, \; n \; even\,. \end{aligned} \tag{219}$$

We demonstrate this for $f=f_1=x^3+x^2+1$ in Figure 16. The convention to go from the ones and zeros of the polynomial and its division to plus and minus ones is $0 \rightarrow +1$ and $1 \rightarrow -1$. [Remember that the coefficients are in GF(2) and thus terms with even coefficients are equivalent to zero.]

$$\begin{array}{r|cccccccccc}
 & 1 & & & x^3 & & x^5 & x^6 & & & \\ \hline
1+x^2+x^3 & 1 & & x^2 & & & & & & & \\
 & 1 & & x^2 & x^3 & & & & & & \\ \cline{2-5}
 & & & & x^3 & & & & & & \\
 & & & & x^3 & & x^5 & x^6 & & & \\ \cline{5-8}
 & & & & & & x^5 & x^6 & & & \\
 & & & & & & x^5 & & x^7 & x^8 & \\ \cline{7-10}
 & & & & & & & x^6 & x^7 & x^8 & \\
 & & & & & & & x^6 & & x^8 & x^9 \\ \cline{8-11}
 & & & & & & & & x^7 & & x^9
\end{array}$$

Figure 16 Calculation of the Characteristic Sequence for $f=x^3+x^2+1$

The Hadamard transform of an m-sequence starting at the beginning of its characteristic sequence or phase has a nice and obvious property. The coefficients of the transform whose positions are at non-zero powers of two will all be equal. This is demonstrated in (220) for the two primitive polynomials of degree 3.

$$
\begin{aligned}
&H_3 \begin{pmatrix} -1 \\ +1 \\ +1 \\ -1 \\ +1 \\ -1 \\ -1 \\ -1 \end{pmatrix} = \begin{pmatrix} -2 & \\ +2 & \leftarrow \\ +2 & \leftarrow \\ -2 & \\ +2 & \leftarrow \\ -2 & \\ -2 & \\ -6 & \end{pmatrix} \qquad (x^3 + x^2 + 1) \\
&H_3 \begin{pmatrix} -1 \\ -1 \\ -1 \\ +1 \\ -1 \\ +1 \\ +1 \\ -1 \end{pmatrix} = \begin{pmatrix} -2 & \\ -2 & \leftarrow \\ -2 & \leftarrow \\ -2 & \\ -2 & \leftarrow \\ -2 & \\ -2 & \\ +6 & \end{pmatrix} \qquad (x^3 + x + 1) \; .
\end{aligned}
\tag{220}
$$

Use of this property was suggested by Yarlagadda and Hershey (1982b) for an FIR filter-based synchronization scheme for m-sequences for which the primitive polynomial was not known beyond its degree.

The Hadamard transform may be used in another way to effect synchronization based on m-sequences. The method is via the "m-sequence transform" which is the cross-correlation of an m-sequence against all $2^n - 1$ phases (and the all zero, degenerate, phase). Because an m-sequence is noise-like, the cross-correlation corresponding to the correct phase will exhibit a large value with respect to the cross-correlation against a different phase even if some of the bits have been corrupted in transmission over a noisy channel. (The number of corrupted bits must, of course, be very much smaller than $2^n - 1$.)

Clearly, direct computation of the cross-correlation requires on the order of 2^{2n} operations which becomes quite large with even modest n. A significant reduction in this work was suggested by Cohn and Lempel (1977). Their construction appeals to a permutational equivalence between an m-sequence transform and the Hadamard transform. Because there is a fast algorithm for the latter, there exists a fast algorithm for the former.

The permutational equivalence rests on a similarity transform developed by Lempel (1979). To construct the m-sequence transform, M_n of dimension $2^n \times 2^n$, we first define a left circulant matrix, S_c, of dimension $(2^n - 1) \times (2^n - 1)$, whose rows are successive end around shifts of an m-sequence of length $2^n - 1$. As an example, let us pick the m-sequence[11] produced by $x^3 + x^2 + 1$ and let us select the phase 1001011 as the first row. For our example, then we have

$$S_c = \begin{pmatrix} 1 & 0 & 0 & 1 & 0 & 1 & 1 \\ 0 & 0 & 1 & 0 & 1 & 1 & 1 \\ 0 & 1 & 0 & 1 & 1 & 1 & 0 \\ 1 & 0 & 1 & 1 & 1 & 0 & 0 \\ 0 & 1 & 1 & 1 & 0 & 0 & 1 \\ 1 & 1 & 1 & 0 & 0 & 1 & 0 \\ 1 & 1 & 0 & 0 & 1 & 0 & 1 \end{pmatrix} . \tag{221}$$

The next step is to embed S_c into a larger matrix, S_n, of dimension $2^n \times 2^n$. We do this by the construction

$$S_n = \begin{pmatrix} 0 & 0^T_{2^n-1} \\ 0_{2^n-1} & S_c \end{pmatrix} \tag{222}$$

where 0_{2^n-1} is a column vector of $2^n - 1$ zeros.

The m-sequence transform, M_n, is gotten by performing customary mapping $0 \leftrightarrow +1,\ 1 \leftrightarrow -1$ on the elements of S_n, ie,

$$\begin{matrix} & \begin{pmatrix} 0 & \leftrightarrow & +1 \\ 1 & \leftrightarrow & -1 \end{pmatrix} & \\ S_n & \longleftrightarrow & M_n \, . \end{matrix} \tag{223}$$

For our example we have

$$S_3 = \begin{pmatrix} 0 & 0 & 0 & 0 & 0 & 0 & 0 & 0 \\ 0 & 1 & 0 & 0 & 1 & 0 & 1 & 1 \\ 0 & 0 & 0 & 1 & 0 & 1 & 1 & 1 \\ 0 & 0 & 1 & 0 & 1 & 1 & 1 & 0 \\ 0 & 1 & 0 & 1 & 1 & 1 & 0 & 0 \\ 0 & 0 & 1 & 1 & 1 & 0 & 0 & 1 \\ 0 & 1 & 1 & 1 & 0 & 0 & 1 & 0 \\ 0 & 1 & 1 & 0 & 0 & 1 & 0 & 1 \end{pmatrix} \tag{224}$$

and

$$M_3 = \begin{pmatrix} +1 & +1 & +1 & +1 & +1 & +1 & +1 & +1 \\ +1 & -1 & +1 & +1 & -1 & +1 & -1 & -1 \\ +1 & +1 & +1 & -1 & +1 & -1 & -1 & -1 \\ +1 & +1 & -1 & +1 & -1 & -1 & -1 & +1 \\ +1 & -1 & +1 & -1 & -1 & -1 & +1 & +1 \\ +1 & +1 & -1 & -1 & -1 & +1 & +1 & -1 \\ +1 & -1 & -1 & -1 & +1 & +1 & -1 & +1 \\ +1 & -1 & -1 & +1 & +1 & -1 & +1 & -1 \end{pmatrix} . \tag{225}$$

---

[11] For this stage of the development we need to express the m-sequence in terms of zeros and ones vice ones and minus ones. To do this we use the customary mapping $\begin{matrix} +1 \leftrightarrow 0 \\ -1 \leftrightarrow 1 \end{matrix}$.

As an example of the application of the m-sequence transform, let us select the following phase of the m-sequence produced by $x^3 + x^2 + 1$:

$$X = (+1 \quad -1 \quad +1 \quad -1 \quad -1 \quad -1 \quad +1 \quad +1)^T . \tag{226}$$

Note that X has eight bits. The eighth bit is the restart of the particular phase, ie, it is the repeat of the first bit.

By direct calculation, we find that

$$M_3X = \begin{pmatrix} 0 \\ 0 \\ 0 \\ 0 \\ 8 \\ 0 \\ 0 \\ 0 \end{pmatrix} \begin{matrix} \\ \\ \\ \\ \leftarrow \\ \\ \\ \\ \end{matrix} \tag{227}$$

the largest coefficient is eight. This is of course the case in which the m-sequence is aligned for maximal autocorrelation.

Returning now to the question of a fast algorithm for computing $M_n X$, we first decompose S_n into the form

$$S_n = C_n C_n^T \tag{228}$$

where C_n is of dimension $2^n \times n$. Such a decomposition can be found using a procedure developed by Lempel (1975). The rows of C_n will be distinct, ie, they will contain all n-tuples. We will now require the matrix B_n, the matrix of binary representations of an n-bit counter, introduced at the beginning of Chapter 6. B_n is of the same dimensions as C_n, which also exhibits all n-tuples and therefore we may write

$$C_n = P_n B_n \tag{229}$$

where P_n is a $2^n \times 2^n$ dimension permutation matrix. We can then perform a following similarity transform, to wit

$$\begin{aligned} S_n &= C_n C_n^T \\ &= P_n (B_n B_n^T) P_n^T \end{aligned} \tag{230}$$

and therefore

$$M_n = P_n H_n P_n^T . \tag{231}$$

For our example,

$$B_3 = \begin{pmatrix} 0 & 0 & 0 \\ 0 & 0 & 1 \\ 0 & 1 & 0 \\ 0 & 1 & 1 \\ 1 & 0 & 0 \\ 1 & 0 & 1 \\ 1 & 1 & 0 \\ 1 & 1 & 1 \end{pmatrix} \tag{232}$$

and

$$C_3 = \begin{pmatrix} 0 & 0 & 0 \\ 1 & 1 & 1 \\ 1 & 1 & 0 \\ 0 & 1 & 1 \\ 0 & 0 & 1 \\ 1 & 0 & 1 \\ 0 & 1 & 0 \\ 1 & 0 & 0 \end{pmatrix} \tag{233}$$

and therefore

$$P_3 = \begin{pmatrix} 1 & 0 & 0 & 0 & 0 & 0 & 0 & 0 \\ 0 & 0 & 0 & 0 & 0 & 0 & 0 & 1 \\ 0 & 0 & 0 & 0 & 0 & 0 & 1 & 0 \\ 0 & 0 & 0 & 1 & 0 & 0 & 0 & 0 \\ 0 & 1 & 0 & 0 & 0 & 0 & 0 & 0 \\ 0 & 0 & 0 & 0 & 0 & 1 & 0 & 0 \\ 0 & 0 & 1 & 0 & 0 & 0 & 0 & 0 \\ 0 & 0 & 0 & 0 & 1 & 0 & 0 & 0 \end{pmatrix}. \tag{234}$$

In Table 20 we demonstrate the computation of M_3X by calculating $P_3H_3P_3^TX$ using the fast Hadamard transform using pairwise sums and differences as depicted, where A is given in (74).

Table 20 Calculation of $P_3 H_3 P_3^T X$ Using the Fast Hadamard Transform

X	$P_3^T X$	Fast Hadamard Transform: $AP_3^T X$	$A^2 P_3^T X$	$H_3 P_3^T X$	$P_3 H_3 P_3^T X$	
↓	↓			↓	↓	
+1	+1	0	0	0	0	
-1	-1	0	0	8	0	
+1	+1	0	4	0	0	
-1	-1	0	4	0	0	
-1	+1	2	0	0	8	←
-1	-1	2	0	0	0	
+1	+1	2	0	0	0	
+1	-1	2	0	0	0	

Further overviews of m-sequences can be found in many publications, for example Golomb (1967) and Chapter 8 in Hershey and Yarlagadda (1986).

Chapter 18 Composite Sequences

Synchronization is an essential discipline of electronic communications. It is also one of the most difficult. This can become quite evident when one is attempting to synchronize two identical generators which produce long periodic sequences. This problem crops up in some types of direct sequence spread spectrum communications as well as distance measuring systems.

Consider that we have a long periodic pseudorandom sequence. In order to synchronize a receiver to the received sequence, we could successively posit and test all possible offsets. For a long sequence this could clearly consume a lot of time.[12]

Titsworth (1964) suggested a very clever way of reducing the complexity of the synchronization search by probabilistically combining several "short" sequences of relatively prime periods. His method depends upon a property of the boolean majority function, $MAJ(\cdot)$. This function has an odd number of binary inputs, $b_1, b_2, \ldots, b_{2r+1}$, $b_i \in \{\pm 1\}$, and is defined as:

$$MAJ(b_1, b_2, \ldots, b_{2r+1}) = \begin{cases} +1, & \sum_{i=1}^{2r+1} b_i > 0 \\ -1, & \sum_{i=1}^{2r+1} b_i < 0 \, . \end{cases} \tag{235}$$

Let us examine the Hadamard transform of the majority function of three variables, $r = 1$, as developed in Table 21. Notice that there are three large

Table 21 The Majority Function of Three Variables and Its Hadamard Transform

b_1	b_2	b_3	$MAJ(b_1, b_2, b_3)$	$H_3(MAJ(b_1, b_2, b_3))$	
1	1	1	1	0	
1	1	-1	1	4	$\leftarrow$
1	-1	1	1	4	$\leftarrow$
1	-1	-1	-1	0	
-1	1	1	1	4	$\leftarrow$
-1	1	-1	-1	0	
-1	-1	1	-1	0	
-1	-1	-1	-1	-4	

[12] A general discussion of this synchronization question is found in Chapter 11 of the text by Hershey and Yarlagadda (1986).

coefficients, indicated by arrows, corresponding to input vectors where only a single bit is a one. This indicates that there is a high correlation of the majority function with each of its individual inputs. Indeed, we see that each b_i agrees with the output of the majority function with probability $\frac{3}{4}$.

Titsworth (1964) shows that for a $2r+1$ input majority function, each b_i will agree with the output of the majority function with probability $\frac{1}{2} + \frac{1}{2^{2r+1}}\binom{2r}{r}$. Titsworth proposed using this property of the majority function in the following way. We construct $2r+1$ *components* or sequences of relatively prime lengths or periods. Let the periods be denoted by $p_1, p_2, \ldots p_{2r+1}$ with $(p_i, p_j) = 1$ for $i \neq j$.[13] We then form the sequence resulting from the output of the majority function whose $2r+1$ inputs are the repeating or cycling sequences. The resulting sequence, the consecutive outputs of the majority function will have a period equal to the product of the individual periods of the input components. The phase of the output sequence is equivalently determined by determining the phases of the components. These can be discovered by cross correlating just a segment of the output sequence with each phase of each component. Once the component sequence phases are discovered, the Chinese Remainder Theorem can be employed to find the number of steps away from any desired epoch. Thus, the synchronization problem is reduced from complexity $O\left(\prod_{i=1}^{2r+1} p_i\right)$ to $O\left(\sum_{i=1}^{2r+1} p_i\right)$.

As an example, consider that $r = 1$ and that the three sequences are denoted as s_1, s_2, and s_3 and are

$$\begin{aligned} s_1 &= +1 \;\; -1 \\ s_2 &= +1 \;\; +1 \;\; -1 \\ s_3 &= +1 \;\; +1 \;\; -1 \;\; -1 \;\; -1\,. \end{aligned} \tag{236}$$

The sequence, S, produced by cycling the three components started at the phases shown in (236) has period $2 \times 3 \times 5 = 30$. From the first three bits, using majority logic, we have $(1,1,1) \rightarrow 1$, second three bits $(-1,1,1) \rightarrow 1$, third three bits $(1,-1,-1) \rightarrow -1$ and so on. Let us consider the first eight bits of this sequence, S_{1-8}. They are $+1\ +1\ -1\ -1\ +1\ -1\ +1\ -1$. We may now recover the phases of the components by performing ten cross-correlations, two for the first sequence, three for the second sequence, and five for the third sequence. The first two of these cross-correlations are the cross-correlations of the two phases of s_1 with S_{1-8}. We illustrate these cross-correlations in Figure 17. We use an 'A' to indicate agreement and a 'D' to indicate a disagreement.

13 $(p_i, p_j) = 1$ stipulates that p_i and p_j are relatively prime as defined earlier in Footnote 10. There are additional considerations on selecting the sequences such as the general cross-correlation between the sequences and the approximate lengths of the sequences. The interested reader is referred to the original paper.

+1	+1	-1	-1	+1	-1	+1	-1	S_{1-8}
+1	-1	+1	-1	+1	-1	+1	-1	**First phase of** s_1
A	D	D	A	A	A	A	A	

+1	+1	-1	-1	+1	-1	+1	-1	S_{1-8}
-1	+1	-1	+1	-1	+1	-1	+1	**Second phase of** s_1
D	A	A	D	D	D	D	D	

Figure 17 Cross-correlations of the Two Phases of s_1 with S_{1-8}

The greatest number of agreement is for the first phase and this indicates that the correct phasing for s_1 is the first phase. Thus, the phase of the first sequence has been recovered. The phases for the remaining two sequences can be recovered in a similar manner. We have thus reduced the work of testing the 30 different phases of the composite sequence to the 10 phases of the component sequences.

Chapter 19 The Thue-Morse Sequence

The Thue-Morse sequence may be defined in a number of ways (Hershey, 1979). Let us look first at a direct construction. Suppose we have a regular binary counting counter in binary, using plus and minus ones and started from zero. The Thue-Morse sequence, Θ, is the product of the counter stage contents as shown in Figure 18.

Counter Stage Contents

...	b_3	b_2	b_1	b_0	Θ
				+1	+1
				-1	-1
			-1	+1	-1
			-1	-1	+1
		-1	+1	+1	-1
		-1	+1	-1	+1
		-1	-1	+1	+1
		-1	-1	-1	-1
	-1	+1	+1	+1	-1
...	⋮	⋮	⋮	⋮	⋮

Figure 18 Direct Construction of the Thue-Morse Sequence

By formula, if we express a non negative integer, k, as

$$k = \left(\frac{1-b_0}{2}\right) \cdot 2^0 + \left(\frac{1-b_1}{2}\right) \cdot 2^1 + \left(\frac{1-b_2}{2}\right) \cdot 2^2 + \cdots \tag{237}$$

where $b_i \in \{\pm 1\}$, then

$$\Theta(k) = \prod_i b_i \,. \tag{238}$$

There are three properties to $\Theta(k)$, $k = 0, 1, 2, \ldots$, that are immediately visible and interesting. First, the sequence never exhibits runs greater than length two. Second, the sequence never becomes periodic. This is easily argued by contradiction.

Assume the sequence does indeed eventually become periodic. There are then two distinct numbers, a and b such that $\Theta(a+i) = \Theta(b+i)$ for $i \geq 0$. Without loss of generality, pick $a > b$. Now choose $i = I$ such that $a + I$ is a power of two. Clearly then $\Theta(a+I) = -1$ and $\Theta(b+I) = -1$ by assumption. Now consider $i = a + 2I$. Clearly $\Theta(2a+2I) = -1$ and $\Theta(b+a+2I) = +1$ which contradicts the assumption that the sequence has become periodic.

The third property is the functional equation that for any k, a non negative integer,

$$\Theta(2^k i) = \Theta(i) \; , \; i \geq 0. \tag{239}$$

We can construct the Thue-Morse sequence by a set of rewriting rules operating on a string. We start with the initial one element string "+1." We then replace the string with a new string formed by a scanning the old string left to right and rewriting a $+1$ with $+1 \quad -1$ and a -1 with $-1 \quad +1$. The Thue-Morse sequence is generated as the rewritings proceed as demonstrated in Figure 19.

+1

$\downarrow$ $\searrow$

+1 -1

$\downarrow$ $\searrow$ $\downarrow$ $\searrow$

+1 -1 -1 +1

$\downarrow$ $\searrow$ $\downarrow$ $\searrow$ $\downarrow$ $\searrow$ $\downarrow$ $\searrow$

+1 -1 -1 +1 -1 +1 +1 -1

$\vdots$ $\vdots$ $\vdots$ $\vdots$ $\vdots$ $\vdots$ $\vdots$ $\vdots$

Figure 19 Generation of the Thue-Morse Sequence by Rewriting

Yet another direct construction is to start with the initial one element string "+1" and continue to right append the inverse of the string to the string. The Thue-Morse sequence is generated as the right appendings take place as depicted in Figure 20.

+1

+1 -1

+1 -1 -1 +1

+1 -1 -1 +1 -1 +1 +1 -1

$\vdots$ $\vdots$ $\vdots$ $\vdots$ $\vdots$ $\vdots$ $\vdots$ $\vdots$ $\cdots$

Figure 20 Generation of the Thue-Morse Sequence by Successive Right-Appending

Still another way to generate the Thue-Morse sequence is via the generating function, $H(z)$ (Yarlagadda and Hershey, 1984) where

$$H(z) = \prod_{i=0}^{\infty} \left(1 - z^{-2^i}\right) . \tag{240}$$

The Thue-Morse sequence will be found to be the coefficients of $z^0 \quad z^{-1} \quad z^{-2} \quad \ldots$.

Finally, we remark that the Thue-Morse sequence is the diagonal sequence, from top left to "bottom" right, of $\lim_{n \to \infty} H_n$, ie, the left-to-right diagonal elements of the ever developing Hadamard matrix.

The power spectrum, $S(\omega)$ is obtained from $H(z)H(z^{-1})$. Now, let

$$S(z) = \prod_{i=0}^{\infty} \left(1 - \frac{1}{2} z^{-2^i} - \frac{1}{2} z^{2^i}\right) \tag{241}$$

and the power spectral density is $S(\omega) = S(z)|_{z=e^{j\omega}}$. That is,

$$S(\omega) = (1 - cos\omega) \prod_{i=1}^{\infty} 2 \cdot sin^2(2^{i-1}\omega) . \tag{242}$$

This form of the power spectral density is a very interesting function. There is no closed form expression for (242) but we can do some elementary analysis.

First, let us define

$$S_{2^k}(\omega) = (1 - cos\omega) \prod_{i=1}^{k} 2 \cdot sin^2(2^{i-1}\omega) . \tag{243}$$

and

$$S(\omega) = \lim_{k \to \infty} S_{2^k}(\omega) . \tag{244}$$

Equation (243) can also be expressed as

$$\begin{aligned} S_{2^k}(\omega) &= \prod_{i=1}^{k} (1 - cos 2^i \omega) \\ &= 1 + \sum_{j=1}^{2^k - 1} a_j cos(j\omega) \end{aligned} \tag{245}$$

where the $\{a_j\}$ are constants. It is clear that

$$\begin{aligned} \frac{1}{2\pi} \int_{-\pi}^{+\pi} S_{2^k}(\omega) d\omega &= \frac{1}{2\pi} \int_{-\pi}^{+\pi} S_{2^{k-1}}(\omega) d\omega \\ &= 1 \end{aligned} \tag{246}$$

which confirms that the energy is invariant over k.

The functions $S_{2^k}(\omega)$, $k = 1,2,3,4$ are plotted in Figure 21. From this figure we can observe the increase in the number of local maxima as we progress from $S_{2^{k-1}}(\omega)$ to $S_{2^k}(\omega)$ and also the decrease in the width of the spectral maxima as k increases.

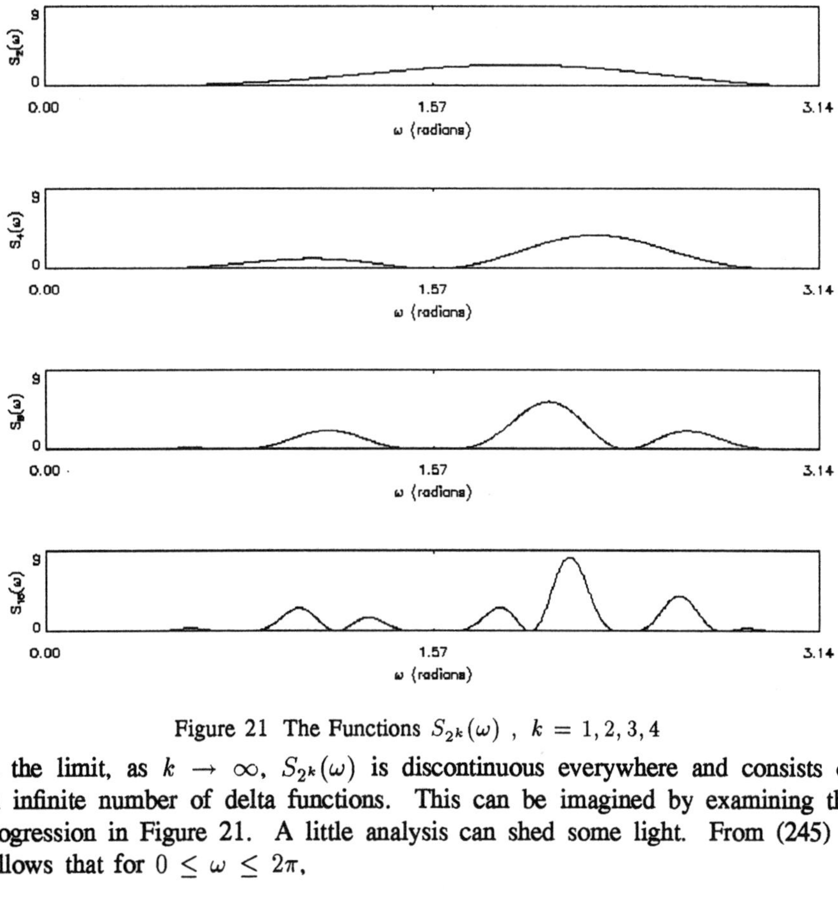

Figure 21 The Functions $S_{2^k}(\omega)$, $k = 1,2,3,4$

In the limit, as $k \to \infty$, $S_{2^k}(\omega)$ is discontinuous everywhere and consists of an infinite number of delta functions. This can be imagined by examining the progression in Figure 21. A little analysis can shed some light. From (245) it follows that for $0 \le \omega \le 2\pi$,

$$S_{2^k}(\omega) = 0 \ , \ \omega = \frac{\pi}{2^k} l \ and \ l = 0, 1, ..., 2^{k+1} \tag{247}$$

and

$$S_{2^k}(\omega) \ne 0 \ , \ for \ l\Big(\frac{\pi}{2^k}\Big) < \omega < (l+1)\Big(\frac{\pi}{2^k}\Big) \tag{248}$$

which indicates that $S_{2^k}(\omega)$ has twice as many zeros as $S_{2^{k-1}}(\omega)$. Further, each maximum in $S_{2^{k-1}}(\omega)$ "splits" into two maxima. In the limit, $S(\omega) = 0$ for $\omega = \frac{n}{2^k}\pi$ for all n and k. There are thus an infinite number of zeros the inter-zero spacings arbitrarily small.

We can examine the behavior of $S(\omega)$ at rational multiples of π by the following argument, given through examples. Let

$$\omega_r = \left(\frac{a}{b}\right)\pi \tag{249}$$

where a and b are integers, $b \neq 0$. To determine if $S(\omega_r)$ is a zero or a delta function we determine the cycle structure of $2^k mod(2\pi b)$ and then determine whether $\prod 2 \cdot sin^2\left[2^i\left(\frac{a}{b}\right)\pi\right]$, computed over a period of the cycle, is less than unity (a zero) or greater than unity (a delta function).

For example, consider $\omega_r = \frac{\pi}{3}$. We calculate as shown in Figure 22.

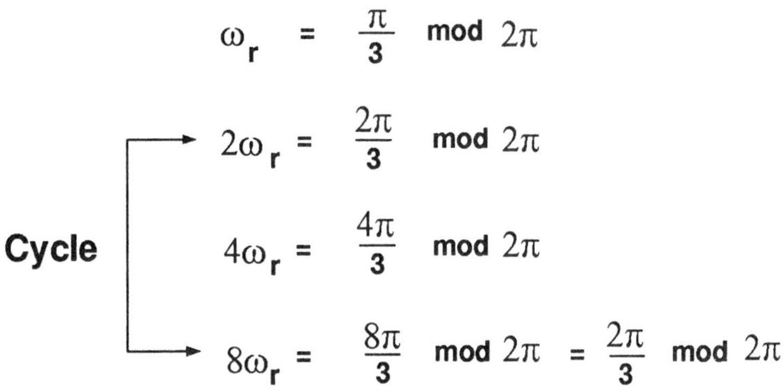

Figure 22 Determination of Cycle Structure for $\omega_r = \frac{\pi}{3}$

We discover that $\omega_r = \frac{\pi}{3}$ is a tail or transient state leading into a two state cycle. To see if $\omega_r = \frac{\pi}{3}$ gives rise to a zero or a delta function we compute

$$\begin{aligned} 2 \cdot sin^2\left(\frac{2\pi}{3}\right) \cdot 2 \cdot sin^2\left(\frac{4\pi}{3}\right) &= (1.5) \cdot (1.5) \\ &= 2.25 \, . \end{aligned} \tag{250}$$

Because the product exceeds unity, we conclude that there is indeed a delta function at $\omega_r = \frac{\pi}{3}$ (and also at $\omega_r = \frac{2\pi}{3}$ and $\omega_r = \frac{4\pi}{3}$ by implication). One can observe these particular delta functions developing by examining the graphs in Figure 21.

Consider now

$$\begin{aligned} S(z) &= \prod_{i=0}^{\infty}\left(1 - \frac{1}{2}z^{-2^i} - \frac{1}{2}z^{2^i}\right) \\ &= \sum_{k=-\infty}^{\infty} R_k z^{-k} \end{aligned} \tag{251}$$

We can obtain the autocorrelation coefficients, $\{R_k\}$ from (251) by computing the inverse transform of $S(z)$. First, $R_0 = 1$ and $R_k = R_{-k}$. The computation of the other coefficients is not so straightforward, however. To start with, let us evaluateR_1. We will consider the first $(k+1)$ and $(k+2)$ products in (251) and then recursively relate the coefficients of $(z^{-1}+z)$ in the two products. Considering, then, the$(k+1)$ products in (251) and expanding, we have

$$\begin{aligned} S_{2^k}(z) &= \prod_{i=0}^{k}\left(1-\frac{1}{2}z^{-2^i}-\frac{1}{2}z^{2^i}\right) \\ &=1+S_{2^k,1}(z^{-1}+z)+F_{2^k}(z) \\ &\quad+(-1)^{k+1}\left(\frac{1}{2}\right)^{k+1}\left(z^{-(2^{k+1}-1)}+z^{(2^{k+1}-1)}\right),\ k>0 \end{aligned} \tag{252}$$

where we have explicitly identified four terms involving $z^0, (z^{-1}+z), \left(z^{-(2^{k+1}-1)}+z^{(2^{k+1}-1)}\right)$, and the terms involving the remaining powers of z identified here by $F_{2^k}(z)$. The first coefficient of $(z^{-1}+z)$ is identified by $S_{2^k,1}$, where the first subscript identifies the subscript of the function $S_{2^k}(z)$ and the second subscript indicates that the coefficient corresponds to the coefficient of $(z^{-1}+z)$.

Forming $S_{2^k}(z)$ we note that

$$\begin{aligned} S_{2^{k+1}}(z) &= S_{2^k}(z)\left(1-\frac{1}{2}z^{-2^{k+1}}-\frac{1}{2}z^{2^{k+1}}\right) \\ &=1+\left(S_{2^k,1}+(-1)^{k+2}\left(\frac{1}{2}\right)^{k+2}\right)(z^{-1}+z) \\ &\quad+F_{2^{k+1}}(z)+(-1)^{k+2}\left(\frac{1}{2}\right)^{k+2}\left(z^{-(2^{k+2}-1)}+z^{(2^{k+2}-1)}\right). \end{aligned} \tag{253}$$

The preceding implies the recursion

$$S_{2^{k+1},1}=S_{2^k,1}+(-1)^{k+2}\left(\frac{1}{2}\right)^{k+2} \tag{254}$$

with $S_{1,1}=-\frac{1}{2}$. Thus, we can determine R_1 as

$$R_1 = \lim_{i\to\infty} S_{2^i,1}=-\frac{1}{2}+\frac{1}{4}-\frac{1}{8}+\frac{1}{16}-\cdots=-\frac{1}{3}. \tag{255}$$

Now consider the product

$$\begin{aligned} B_{2^j}(z) &= \prod_{i=j}^{\infty}\left(1-\frac{1}{2}z^{-2^i}-\frac{1}{2}z^{2^i}\right) \\ &= 1+\sum_{k=1}^{\infty}R_{k\cdot 2^j}\left(z^{-k\cdot 2^j}+z^{k\cdot 2^j}\right) \end{aligned} \tag{256}$$

which exhibits only integral powers of $z^{k\cdot 2^j}$. Also consider

$$
\begin{aligned}
B_{2^{j-1}}(z) &= \left(1 - \frac{1}{2}z^{-2^{j-1}} - \frac{1}{2}z^{2^{j-1}}\right) B_{2^j}(z) \\
&= R_0 + \sum_{k=1}^{\infty} R_{k\cdot 2^{j-1}} \left(z^{-k\cdot 2^{j-1}} + z^{k\cdot 2^{j-1}}\right)
\end{aligned}
\tag{257}
$$

where

$$
\begin{aligned}
&R_0 = 1 \\
&R_{2k\cdot 2^{j-1}} = R_{k\cdot 2^j} \\
&R_{2^{j-1}} = -\frac{1}{2}(R_{2^j} + 1) \\
&R_{(2k-1)\cdot 2^{j-1}} = -\frac{1}{2}\left(R_{(k-1)\cdot 2^j} + R_{k\cdot 2^j}\right) .
\end{aligned}
\tag{258}
$$

Now consider the last equation in (258) with $R_1 = -\frac{1}{3}$ and $k = 1$, ie, $R_{2^j} = -2R_{2^{j-1}} - 1$. By successive substitution we find that

$$
R_{2^j} = -\frac{1}{3}
\tag{259}
$$

implying that $R_{2^j} = R_{2^{j+1}}$. Furthermore, we note from (256) that $B_{2^j}(z) = S_{2^j}(z)$ and it then follows that

$$
R_{k\cdot 2^j} = R_k \ .
\tag{260}
$$

Note that the autocorrelation comports to the same functional equation (239) as the Thue-Morse sequence itself.

The above properties produce an interesting symmetry in the sequence

$$
R_{2^j}, R_{2^j+1}, \ldots, R_{2^{j+1}} \ .
\tag{261}
$$

It is clear from (258) that

$$
R_{2^j+1}, \ldots, R_{2^j+2^{j-1}-1}
\tag{262}
$$

are generated from $R_{2^j} = -\frac{1}{3}$ and $R_{2^j+2^{j-1}} = \frac{1}{3}$ and similarly

$$
R_{2^j+2^{j-1}+1}, \ldots, R_{2^{j+1}-1}
\tag{263}
$$

are generated from $R_{2^j+2^{j-1}} = \frac{1}{3}$ and $R_{2^{j+1}} = -\frac{1}{3}$ in exactly the same manner. Now, because the starting points or boundary conditions are the same, ie, $\left(-\frac{1}{3}, \frac{1}{3}\right)$, and the equations that are used to compute (262) and (263), the sequence (263) will have an even symmetry with center at $R_{2^j+2^{j-1}}$. In other terms,

$$
R_{2^j+k} = R_{2^{j+1}-k} \ .
\tag{264}
$$

Now consider the symmetry aspects of the sequence

$$R_{2^j}, R_{2^j+1}, ..., R_{2^j+2^{j-1}} \ . \tag{265}$$

We use the same reasoning as before but we start with the boundary conditions $\left(-\frac{1}{3}, \frac{1}{3}\right)$ rather than the boundary conditions $\left(-\frac{1}{3}, -\frac{1}{3}\right)$ which were used before. First, we not that $R_{2^j+2^{j-2}} = 0$. We therefore use $\left(-\frac{1}{3}, 0\right)$ as boundary conditions for evaluating the first half of (262). To evaluate the latter half of (262) we use $\left(0, \frac{1}{3}\right)$. This implies an odd symmetry in (265).

The foregoing symmetry aspects imply that

$$\sum_{k=0}^{2^j} R_{2^j+k} = -\frac{1}{3} \ . \tag{266}$$

A short table of $\{R_i\}$ is given in Table 22. The values $\{R_i\}$ are limit values, ie, values produced by averaging an infinite number of terms. As discussed in Yarlagadda and Hershey (1990), reasonably accurate estimates, $\left\{\hat{R}_i\right\}$, may be gotten by averaging only a relatively short segment of the Thue-Morse sequence autocorrelation coefficients.

Table 22 A Short Table of $\{R_i\}$

i	R_i	i	R_i	i	R_i	i	R_i
0	1	8	$-\frac{1}{3}$	16	$-\frac{1}{3}$	24	$\frac{1}{3}$
1	$-\frac{1}{3}$	9	$\frac{1}{6}$	17	$\frac{1}{12}$	25	$-\frac{1}{12}$
2	$-\frac{1}{3}$	10	0	18	$\frac{1}{6}$	26	$-\frac{1}{6}$
3	$\frac{1}{3}$	11	$-\frac{1}{6}$	19	$-\frac{1}{12}$	27	$\frac{1}{12}$
4	$-\frac{1}{3}$	12	$\frac{1}{3}$	20	0	28	0
5	0	13	$-\frac{1}{6}$	21	$\frac{1}{12}$	29	$-\frac{1}{12}$
6	$\frac{1}{3}$	14	0	22	$-\frac{1}{6}$	30	$\frac{1}{6}$
7	0	15	$\frac{1}{6}$	23	$-\frac{1}{12}$	31	$\frac{1}{12}$

A T-point long segment of the TMS sequence, ie $\Theta(i)$, $k \leq i < (k + T - 1)$, can be used to transmit a small amount of information through a high noise environment by having the receiver solve for the transmitter picked value of k [See Yarlagadda and Hershey (1990)]. This is essentially a synchronization scheme.

Chapter 20 Signal Representation

We present a new method for representing and approximating signals which are made up of odd frequency components. The method uses a "half basis" set, all of whose members exhibit identical power spectral densities. In contrast to the conventional Fourier approximation, approximations by this new method will spread the approximation error over the odd frequencies in the half basis. Approximations to the full set of signals, ie, those containing both even and odd frequency components can also be realized by this new method.

It is customary to speak of a partition of the components of a space or time function as *even* or *odd*. It is equally reasonable to posit a partition of the components of a function into parts that comprise only even frequencies and those that comprise only odd frequencies. The characteristic of an odd frequency function, f(x), over the interval [0,1] is $f(x+0.5) = -f(x)\ ,\quad 0 \le x \le 0.5$. Likewise the characteristic of an even frequency function over the same interval is $f(x+0.5) = +f(x)\ ,\quad 0 \le x \le 0.5$.

An example of an even frequency function is $f(x) = sin(4\pi x)$. An example of an odd frequency function is $f(x) = sin(2\pi x)$.

The half-basis grows naturally out of a study of a particular set of functions which are defined over an interval [0,1] by the following expression.

$$W_n(\Phi; x) = \prod_{i=0}^{n} sin(2^i \cdot 2\pi x + \phi_i) \tag{267}$$

where n is the "order" of the function, $\Phi = (\phi_0, \phi_1, ..., \phi_n)$, and $\phi_i \epsilon \{0, \frac{\pi}{2}\}$.

By writing

$$sin(2^i \cdot 2\pi x + \phi_i) = \frac{1}{2j}\left[e^{j(2^i \cdot 2\pi x + \phi_i)} - e^{-j(2^i \cdot 2\pi x + \phi_i)}\right] \tag{268}$$

it is easily seen that the form (267) can be expanded into a sum form as

$$W_n(\Phi; x) = 2^{-n} \sum_{k=1,3,5,...}^{2^{n+1}-1} \Theta\left(\frac{k-1}{2}\right)\begin{cases}(-1)^{n/2} sin\,(k \cdot 2\pi x + \mathrm{A}\bullet\Phi)\ ,\ n,\ even \\ (-1)^{(n-1)/2} cos\,(k \cdot 2\pi x + \mathrm{A}\bullet\Phi)\ ,\ n,\ odd\end{cases} \tag{269}$$

where

i. $\mathrm{A} = (\alpha_0, \alpha_1, ..., \alpha_n)$ is a binary representation of the summation index k with $\alpha_i \epsilon \{\pm 1\}$ and $k = \sum_{i=0}^{n} \alpha_i \cdot 2^i$,

ii. $\Phi = (\phi_0, \phi_1, ..., \phi_n)$,

iii. $\mathrm{A} \bullet \Phi = \sum_{i=0}^{n} \alpha_i \phi_i$,

iv. $\Theta(k),\ k = 0, 1, 2, \ldots$ is the Thue-Morse sequence.

For finite order, n, the functions and all of their derivatives are continuous in the interval (0,1). In Figure 23 we plot some representatives of $2^{\frac{n+1}{2}} W_n(\Phi; x)$, namely $2^{\frac{1}{2}} W_0(\underline{0}; x)$, $2^2 W_3(\underline{0}; x)$, and $2^{\frac{7}{2}} W_6(\underline{0}; x)$, where $\underline{0}$ indicates that all of the ϕ_i's are zero.

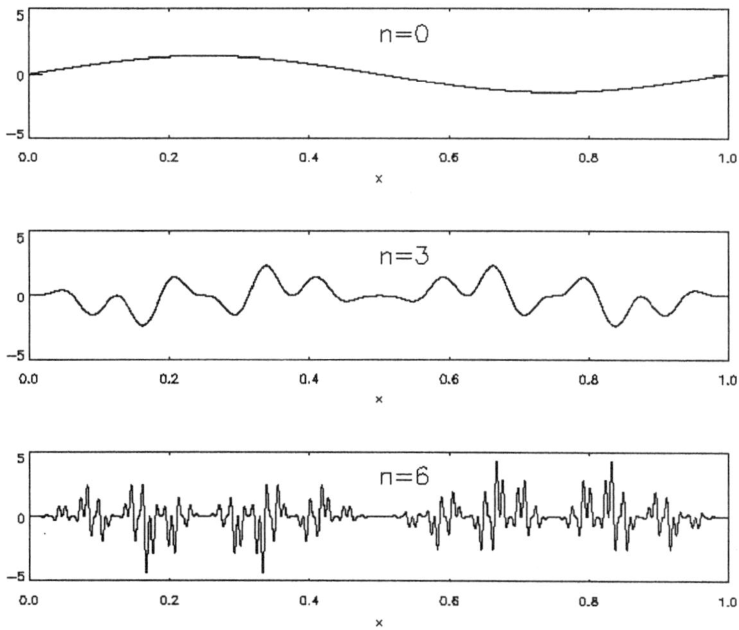

Figure 23 Graphs of $2^{\frac{n+1}{2}} W_n(\underline{0}; x)$ for n=0,3,6

The function set $\{W_n(\Phi; x)\}$ has a number of properties worth noting. First, by directly integrating, it is evident that the functions of order n form an orthogonal set, ie,

$$\int_0^1 W_n(\Phi_1; x) W_n(\Phi_2; x) dx = \begin{cases} 0, & \Phi_1 \neq \Phi_2 \\ 2^{-(n+1)}, & \Phi_1 = \Phi_2 \end{cases}. \tag{270}$$

Second, from (269) we see that the power spectrum of $W_n(\Phi; x)$ is bi-valued. It has a value of $2^{-(2n+1)}$ at the odd frequencies, 1 through $2^{n+1} - 1$, and is zero at the even frequencies.

Third, the autocorrelation, $R_{W_n, W_n}(\tau)$, of a member of $\{W_n(\Phi; x)\}$ will exhibit a very sharp peak about $\tau = 0$. This can be seen by noting that the autocorrelation at $\tau = \frac{1}{2^{n+2}}$ is approximately zero for moderate n. This is clear from the definition of the autocorrelation

$$R_{W_n, W_n}(\tau) = \int_0^{1-\tau} W_n(\Phi_1; \acute{x})|_x W_n(\Phi_1; \acute{x})|_{x+\tau} dx \tag{271}$$

where $\Phi_1 = (\phi_0, \phi_1, ..., \phi_n)$ and $W_n(\Phi_1; \acute{x})|_\nu$ specifies the evaluation of $W_n(\Phi_1; \acute{x})$ at $\acute{x} = \nu$. When $\tau = \frac{1}{2^{n+2}}$ we can express the term $W(\Phi_1; \acute{x})|_{x+\tau}$ as

$$W_n(\Phi_1; \acute{x})|_{x+\frac{1}{2^{n+2}}} = sin\Big[2^n \cdot 2\pi x + \Big(\phi_n + \frac{\pi}{2}\Big)\Big] \sum_{\Phi} \alpha_\Phi W_{n-1}(\Phi; \acute{x})|_x \tag{272}$$

where the $\{\alpha_i\}$ are coefficients of functions of order n-1. Clearly $W_n(\Phi_1; \acute{x})|_{x+\frac{1}{2^{n+2}}}$ is orthogonal to $W_n(\Phi_1; \acute{x})|_x$ over the interval [0,1] and thus approximately orthogonal over the diminished interval $\big[0, 1 - \frac{1}{2^{n+2}}\big]$ for moderate n.

The normalized autocorrelations of the three waveforms presented in Figure 23 are plotted in Figure 24.

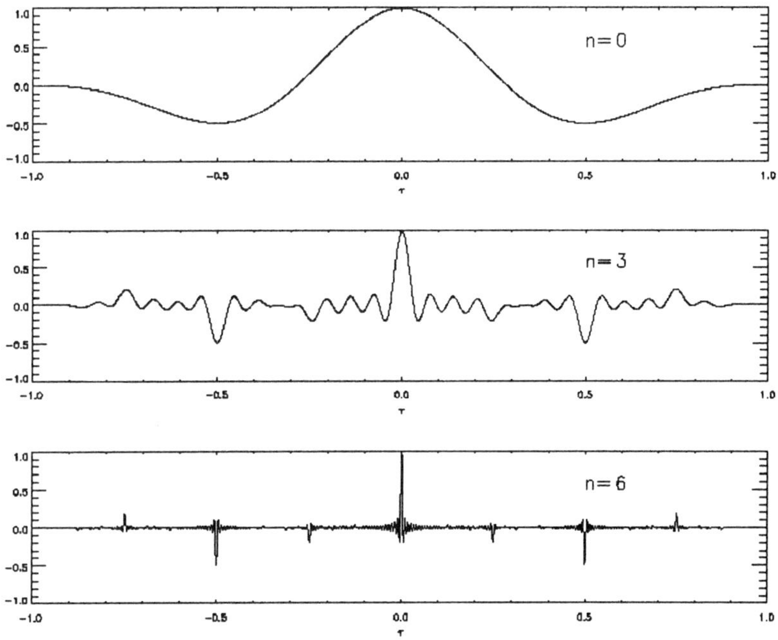

Figure 24 Normalized Autocorrelations of $2^{\frac{n+1}{2}} W_n(\underline{0}; x)$ for n=0,3,6

The fourth property is hierarchical. It is possible to find a set $\{c_\Phi\}$ of order n coefficients, such that

$$\sum_{\Phi} c_\Phi W_n(\Phi; x) = sin(k \cdot 2\pi x) \; or \; cos(k \cdot 2\pi x), \quad k = 1, 3, ..., 2^{n+1} - 1\,. \tag{273}$$

This property may be concluded by expanding

$$sin(k \cdot 2\pi x + A \bullet \Phi) = sin(k \cdot 2\pi x) cos(A \bullet \Phi) + cos(k \cdot 2\pi x) sin(A \bullet \Phi) \tag{274}$$

and noting that the half set of $\{W_n(\Phi; x)\}$ corresponding to the summation indices that have an even number of ones in their binary representations yields a matrix of $\{cos(A \bullet \Phi)\}$ that has full rank. Thus any odd frequency cosine term may be isolated. A similar argument obtains for any odd frequency sine term.

This property allows a convenient multiresolution approximation to obtain wherein any approximation of $\sum_{\Phi} c_{\Phi} W_n(\Phi; x)$ is *"upwardly subsumed"* by, or *embedded in*, the set of possible approximations of order n+1. This property essentially ensures that signal reconstruction based on this method of signal representation will be numerically well behaved.

The fifth property concerns the relationship between the set of functions $\{W_n(\Phi; x)\}$ and a wavelet *eigensequence*. As Strang (1989) discusses, wavelets are developed from first choosing a basis function, $\phi(x)$, and solving for the set of coefficients $\{c_k\}$ which operate in the two scale dilation equation $\phi(x) = \sum c_k \phi(2x - k)$. If we choose $\phi(x)$ to be the "box" function, which is unity on the interval [0,1] and zero elsewhere, we find that $c_0 = 1$ and $c_1 = 1$. The wavelet that results from this selection is the *Haar* wavelet, which is visible on forming $\phi(2x) - \phi(2x - 1)$. The Discrete Haar Wavelet Transform can be easily performed on a set of 2^n points by *hierarchically* performing quadrature mirror filtering as instructed in Press (1992, Section 13.10).

If H is a discrete wavelet transform and s is a data sequence, then an eigen-type relation obtains with a condition such as $Hs = \lambda s$. In general, there is only a trivial solution to this equation in the wavelet domain. If, however, we allow the definition to be modified to a recursive format such as $Hs_{2^n}^T = \lambda(0_{2^{n-1}} : s_{2^{n-1}})^T$ in which s_{2^n} is a 2^n long sequence and $0_{2^{n-1}}$ is a 2^{n-1} long sequence of zeros and the colon denotes concatenation, then we can find some interesting solutions. The solutions will, of course, be intimately related to the particular wavelet. Choosing the Haar wavelet, we can show that the following constitutes such a solution with $\lambda = \frac{1}{2}$. We let s_{2^n} be the first 2^n terms of the Thue-Morse sequence.

If we properly sample $W_n(\underline{0}; x)$ and hard quantize the samples to $\{\pm 1\}$, we will obtain the first 2^{n+1} terms of the Thue-Morse sequence. The proper sampling of the function is at the ever increasing values of x over the interval [0,1] at what we term the *semi-modal points* $\{x_i\}$ where

$$x_i = \frac{i}{2^{n+2}}, \; i = 1, 3, 5, ..., 2^{n+2} - 1. \tag{275}$$

The traces shown in Figure 25 summarize the results so far using $W_6(\underline{0}; x)$ as an example.

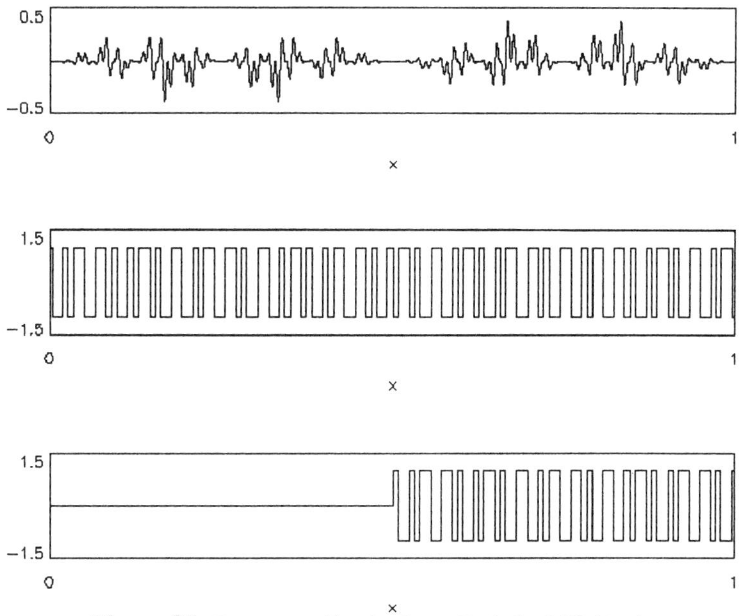

Figure 25 Top trace: Graph of one Period of $W_6(\underline{0}; x)$
Middle trace: Hard Quantization of $W_6(\underline{0}; x)$ Sampled at the Mode Points
Bottom trace: $\frac{1}{2}$·Haar Wavelet Transform of the above Samples

The sixth property is a behavioral similarity to Weierstrass functions. Recently Resnikoff (1990) published an intriguing connection between Weierstrass functions and compactly supported wavelets. Weierstrass functions are of the form $\sum_{i=0}^{\infty} a^i cos b^i \pi x$. This class of functions is interesting in its own right as the functions are everywhere continuous and nowhere differentiable for some values of the defining parameters a and b.

In the limit, as $n \rightarrow \infty$, the function, $2^n W_n(\underline{0}; x)$, remains continuous over the interval (0,1) but becomes non-differentiable within the interval. To see this, consider one of the identities due to Euler given in (Resnikoff 1990) viz

$$\frac{1}{2} = 1 \pm cos\, \nu + cos\, 2\nu \pm cos\, 3\nu + cos\, 4\nu \pm \ldots \qquad (276)$$

for $0 < \nu < \pi$. From this we can see that

$$0 = \pm cos\, \nu \pm cos\, 3\nu \pm cos\, 5\nu \pm cos\, 7\nu \pm \ldots . \qquad (277)$$

As the Thue-Morse sequence, $\Theta(\cdot)$, takes on only the values $\{\pm 1\}$, we conclude that, in the limit, $2^n W_n(\underline{0}; x)$ remains continuous but cannot be differentiated and is thus a Weierstrass-like function. In the limit, the eigensequence becomes an *eigenprocess* which can be set in the framework of the Continuous Haar Wavelet Transform.

Now, the set of functions $\{W_n(\Phi; x)\}$ can be used to represent by approximation an odd frequency function on the interval [0,1]. To do this we form

$$z(x) = \sum_{\Phi} c_{\Phi} W_n(\Phi; x) \tag{278}$$

where

$$c_{\Phi} = 2^{n+1} \int_0^1 z(x) W_n(\Phi; x) dx \,. \tag{279}$$

What is especially interesting about this approach to continuous signal representation is that the basis functions, unlike the elements of the usual bases, all have the same power spectral density as is evident from (269).

Asymptotically, as $n \to \infty$, the functions $\{W_n(\Phi; x)\}$ form a basis capable of representing odd frequency functions. The functions are orthogonal and belong to the Hilbert space of finite energy functions with a norm defined by

$$< z, W_n(\Phi; x) >= 2^{n+1} \int_0^1 z(x) W_n(\Phi; x) dx \,. \tag{280}$$

See, for example, the Fourier Series Theorem in Naylor and Sell (1982, Theorem 5.17.8).

The signal representation can be made to include even functions also, and thus all functions, by a number of means. One such way is simply to decompose the function to recover the odd frequency components, as per (279), and then invert the right half of the function and decompose again for the even frequency components.

As an example of continuous signal representation, we represent by approximation, $z^*(x)$, the odd frequency function

$$z(x) = \begin{cases} +sin(4\pi x), & 0 \leq x < 0.5 \\ -sin(4\pi x), & 0.5 \leq x \leq 1.0 \end{cases} \tag{281}$$

by (278) for order six. The results are shown in Figure 26.

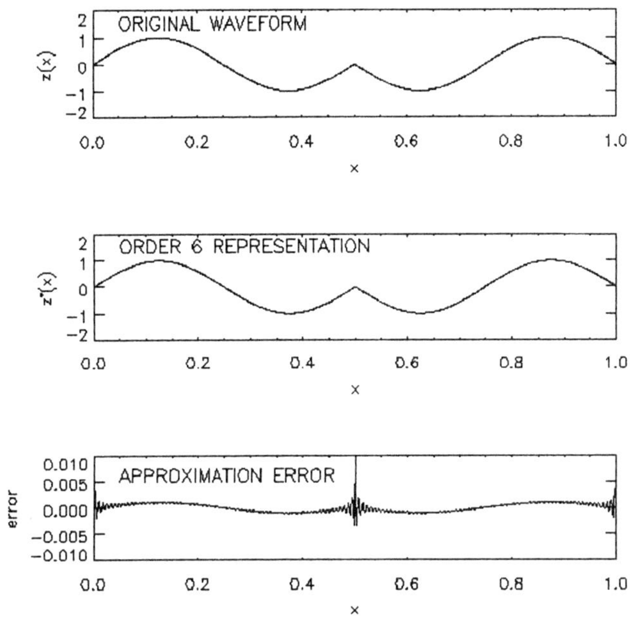

Figure 26 An Odd Function, Its Order 6 Approximation, and the Approximation Error

Chapter 21 A New Hadamard Basis and Its Implication for Signaling

A commonly used technique for achieving nominal spectrum spreading is to use a family of Hadamard or Walsh-Hadamard basis vectors. [See Wilson and Lakshman, (1968).] We can develop a new basis out of the functions presented in (267) that preserves its relation to the Sylvester-Hadamard matrix in sign sequency of its basis vectors but does not consist of $\{\pm 1\}$ elements. This new basis may be well suited for some new signaling and filtering designs.

We form the basis for order n, $B_n(X)$, consisting of 2^n vectors of 2^n values each, each vector, $B_{n,\Phi}(X)$, corresponding to one of the 2^n unique values of Φ. The vector values are obtained for each vector by sampling $W_{n-1}(\Phi; x)$, defined in (267), at the *modal points* $\{X_i\}$ where

$$X_i = \frac{2i+1}{2^{n+2}}\ , \ i = 0, 1, 2, ..., 2^n - 1. \tag{282}$$

It is to be noted that this new basis is orthonormal.

As an example, we display the new basis for n=3. Notice that the vectors are arranged to comport to the sign order found in H_3. (Such an arrangement is possible for general n.)

$$B_3 = \begin{pmatrix} +0.641 & +0.225 & +0.150 & +0.127 & +0.127 & +0.150 & +0.225 & +0.641 \\ +0.641 & -0.225 & +0.150 & -0.127 & +0.127 & -0.150 & +0.225 & -0.641 \\ +0.265 & +0.543 & -0.363 & -0.053 & +0.053 & +0.363 & -0.543 & -0.265 \\ +0.265 & -0.543 & -0.363 & +0.053 & +0.053 & -0.363 & -0.543 & +0.265 \\ +0.127 & +0.150 & +0.225 & +0.641 & -0.641 & -0.225 & -0.150 & -0.127 \\ +0.127 & -0.150 & +0.225 & -0.641 & -0.641 & +0.225 & -0.150 & +0.127 \\ +0.053 & +0.363 & -0.543 & -0.265 & -0.265 & -0.543 & +0.363 & +0.053 \\ +0.053 & -0.363 & -0.543 & +0.265 & -0.265 & +0.543 & +0.363 & -0.053 \end{pmatrix}. \tag{283}$$

In Figure 27 we display the eight 8-point basis vectors. We connect the points with a dotted line so that the vectors can be easily considered as signaling waveforms.

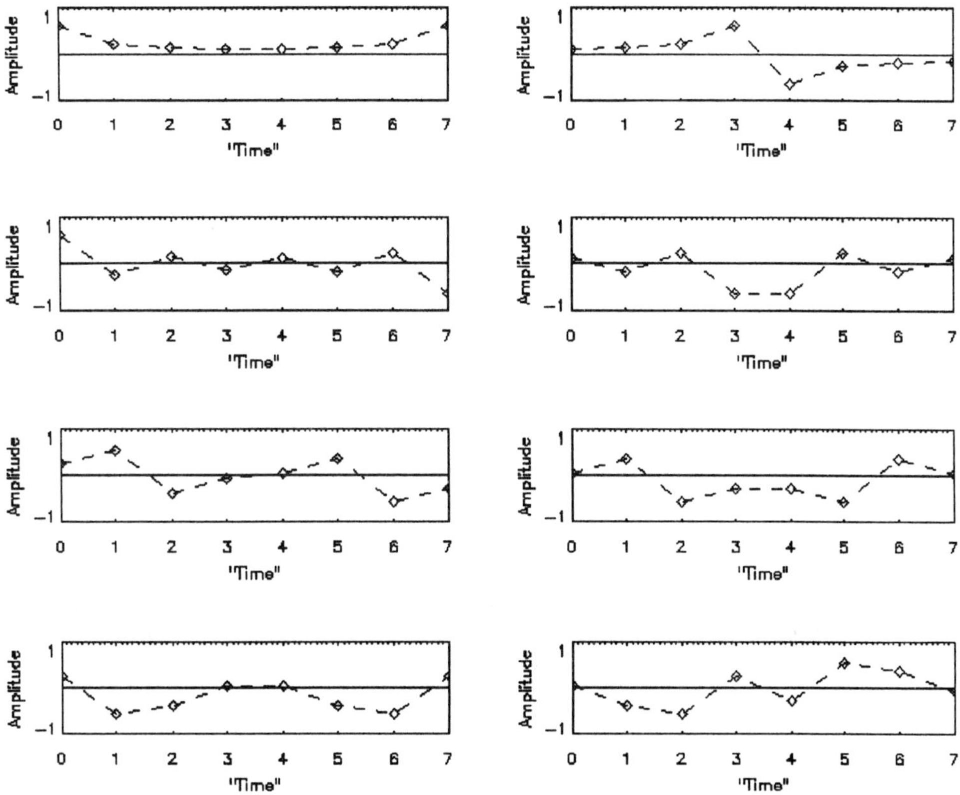

Figure 27 The Eight Vectors of B_3

The sequences from B_n will have some different properties than those from $2^{-\frac{n}{2}}H_n$. The chief contrast between the new basis and the traditional Hadamard basis is a trade-off between the average auto and cross correlations, and the distribution of power spectral density. Figure 28 displays the average absolute value of the auto and crosscorrelations of the individual sequences. The solid lines correspond to the sequences taken from $2^{-\frac{n}{2}}H_n$ and the dotted lines to those sequences taken from B_n. (All auto and cross correlations were averaged *with the exception of* the autocorrelation of lag zero.)

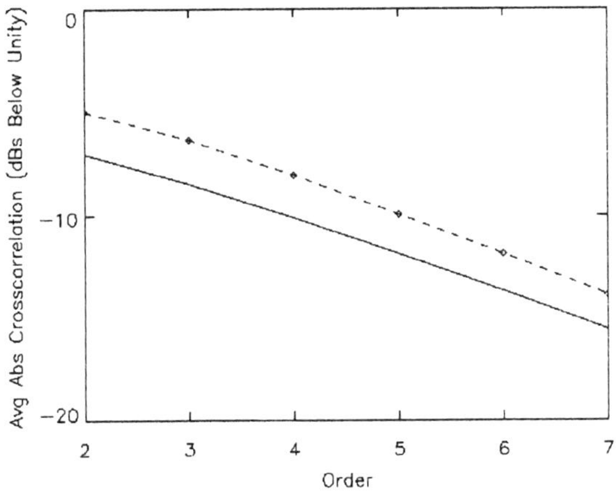

Figure 28 Average Auto and Cross-correlations

Figure 29 displays the power spectral density of the overlaid sign-similar sequences. Again, the solid lines correspond to the sequences $2^{-\frac{n}{2}}H_n$ and the dotted lines to those sequences taken from B_n. (The displays are such that the power densities are appropriately normalized and the display is taken out to the folding frequency.) Notice that the power of the sequences from B_n is spread more across the band than is the power for the sequences from $2^{-\frac{n}{2}}H_n$ which is generally concentrated in one part of the spectrum. This property which accrues from selecting sequences from B_n over those from $2^{-\frac{n}{2}}H_n$ may provide better signal error rates through selective fading and may also allow filters for out of band signal suppression to be less complex and expensive.

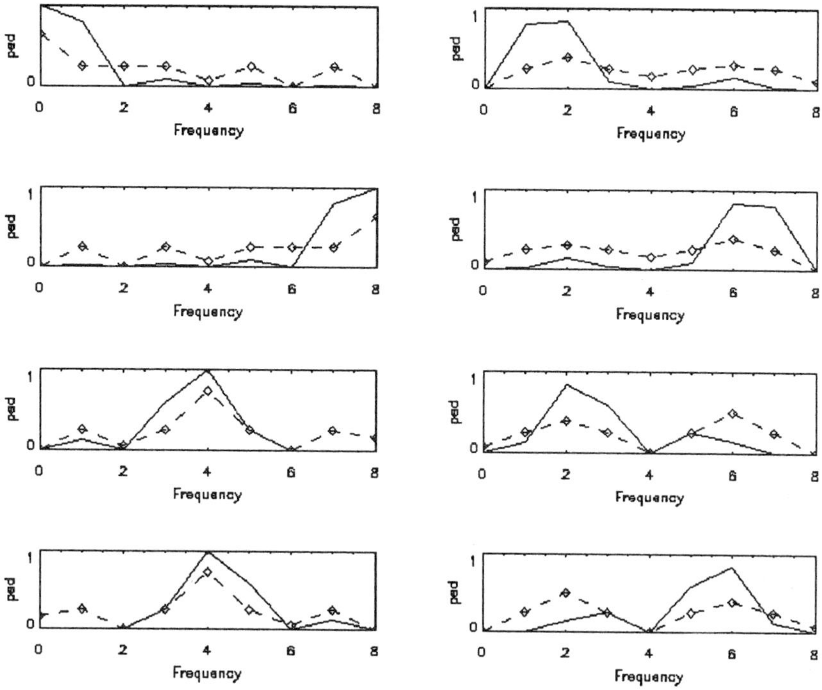

Figure 29 The Power Spectral Density (psd) out to the Folding Frequency and Appropriately Normalizing the Power

Chapter 22 A Stochastic Operator Field

As shown in Yarlagadda and Hershey (1985), if we define I_n as the identity matrix of order n and $\mathrm{H}_n' = 2^{-n/2}H_n$, then we can stochastically construct an operator, $\Psi_n(\alpha)$, as

$$\Psi_n(\alpha) = \alpha I_n + (1-\alpha)\mathrm{H}_n' \, , \tag{284}$$

where H_n is the Hadamard matrix and H_n' is the normalized Hadamard matrix. The following identities are easily established:

$$
\begin{aligned}
&\Psi_n(1) = I \\
&\Psi_n(0) = \mathrm{H}_n' \\
&\Psi_n(\alpha) = \alpha\Psi_n(1) + (1-\alpha)\Psi_n(0) \qquad (a) \\
&\Psi_n(\alpha_1+\alpha_2) = \Psi_n(\alpha_1) + \Psi_n(\alpha_2) - \Psi_n(0) \qquad (b) \\
&\Psi_n(\alpha_1)\Psi_n(\alpha_2) = \Psi_n(\alpha_1\alpha_2 + (1-\alpha_1)(1-\alpha_2)) \qquad (c) \\
&\Psi_n(\alpha)\Psi_n\left(\frac{\alpha}{2\alpha-1}\right) = \Psi_n(1)\ , \ \alpha \neq \frac{1}{2} \ (the\ inverse) \qquad (d) \\
&\Psi_n(\alpha)\Psi_n\left(\frac{\alpha-1}{2\alpha-1}\right) = \Psi_n(0)\ , \ \alpha \neq \frac{1}{2} \qquad (e) \\
&\Psi_n^2(\alpha) = \Psi_n\left(\alpha^2 + (1-\alpha)^2\right) \qquad (f) \\
&\Psi_n^m(\alpha) = \Psi_n\left(\frac{1+(2\alpha-1)^m}{2}\right) \qquad (g) \\
&\Psi_n(\alpha_1\alpha_2) = \frac{\alpha_1\Psi_n(\alpha_2) + \alpha_2\Psi_n(\alpha_1)}{2} + \left(\frac{\alpha_1+\alpha_2}{2} - 1\right)\Psi_n(0) \qquad (h)\ .
\end{aligned}
\tag{285}
$$

The convex form (284) gives rise to many properties. One result is the formation of a field under the operations of common matrix multiplication and a special addition operator denoted by $'\star'$ and defined as

$$\Psi_n(\alpha_1) \star \Psi_n(\alpha_2) = \Psi_n(\alpha_1) + \Psi_n(\alpha_2) - \Psi_n\left(\frac{1}{2}\right) \tag{286}$$

where $'+'$ is normal addition.

Chapter 23 A Naturalness Preserving Transform

The use of (284) has been suggested as the basis for a "naturalness preserving" transform. (J. Hershey and R. Yarlagadda, 1983; R. Yarlagadda and J. Hershey, 1985.) Consider that we have an image encoded as a square matrix, P_n, that consists of $N \times N$ pixels where $N = 2^n$. We form

$$\Pi_n = (\pi_{ij}) = \Psi_n(\alpha) P_n \Psi_n(\alpha) \ . \tag{287}$$

If α is close to unity, then we will still be able to recognize the original image P_n in Π_n. By using (285d) we can recover P_n from Π_n, ie,

$$P_n = \Psi_n\left(\frac{\alpha}{2\alpha - 1}\right) \Pi_n \Psi_n\left(\frac{\alpha}{2\alpha - 1}\right), \ \alpha \neq \frac{1}{2} \ . \tag{288}$$

We can gain some further insight into Π_n by exploring the Euclidean norm of (287). First we note that $\mathrm{H}_n^{'}$ can be written as [See Chapters 4 and 5.]

$$\mathrm{H}_n^{'} = C^T \begin{pmatrix} I_{2^{n-1}} & 0 \\ 0 & -I_{2^{n-1}} \end{pmatrix} C \tag{289}$$

where C is an orthogonal matrix of order 2^n and $I_{2^{n-1}}$ is an identity matrix of order 2^{n-1}.

We write

$$\alpha I_{2^n} + (1 - \alpha)\mathrm{H}_n^{'} = C^T D C \tag{290}$$

where I_{2^n} is an identity matrix of order 2^n and D is the diagonal matrix

$$D = \begin{pmatrix} I_{2^{n-1}} & 0 \\ 0 & (2\alpha - 1) I_{2^{n-1}} \end{pmatrix} . \tag{291}$$

If we now set

$$\Pi_n = \Psi_n(\alpha) P_n \Psi_n(\alpha) \tag{292}$$

we can write

$$\Pi_n = C^T D C P_n C^T D C \tag{293}$$

and we have

$$||\Pi_n|| = \left\| C^T D C P_n C^T D C \right\| . \tag{294}$$

Note that, in general,

$$||CB|| = ||B|| \tag{295}$$

for C an orthogonal matrix. [See Wilkinson (1965), for example.] Therefore,

$$||\Pi_n|| = \left\|DCP_nC^TD\right\| \leq ||D||\left\|CP_nC^T\right\|||D|| \tag{296}$$

and

$$||\Pi_n|| \leq ||D||||P_n||||D|| \tag{297}$$

with

$$||D|| = 2^{\frac{n-1}{2}}\sqrt{\left(1+(2\alpha-1)^2\right)} \tag{298}$$

which is a minimum for $\alpha = \frac{1}{2}$.

Expanding (287), we have[14]

$$\Pi_n = \alpha^2 P_n + (1-\alpha)^2 \mathrm{H}_n^{'} P_n \mathrm{H}_n^{'} + \alpha(1-\alpha)\left(P_n\mathrm{H}_n^{'} + \mathrm{H}_n^{'} P_n\right) . \tag{299}$$

The first term on the right side of (299) is the weighted original image. The second term is the weighted normalized 2–D Hadamard transform. The third term is a sort of cross product that is, incidentally, invariant under $\Psi_n(0)$ and $\Psi_n(1)$, ie,

$$\begin{aligned} P_n\mathrm{H}_n^{'} + \mathrm{H}_n^{'} P_n &= \Psi_n(0)\left(P_n\mathrm{H}_n^{'} + \mathrm{H}_n^{'} P_n\right)\Psi_n(0) \\ &= \Psi_n(1)\left(P_n\mathrm{H}_n^{'} + \mathrm{H}_n^{'} P_n\right)\Psi_n(1) . \end{aligned} \tag{300}$$

The important thing to note is that $\Psi_n(\alpha)P_n\Psi_n(\alpha)$ is in a sense *holographic* as it involves the Hadamard transform, a completely dense matrix, and therefore every element of $\Psi_n(\alpha)P_n\Psi_n(\alpha)$ is a function of the entire image.

For a significant range of α below unity, the sense of the scene is interpretable, the features are recognizable, but there is a large Hadamard component. Thus, assume a section of Π_n is lost or discarded. We can attempt to recover all of P_n by positing the contents of the lost or discarded portion, ie, by filling in the missing region with a candidate subscene and then inverse transforming. If that part of P_n which we could see before inverse transforming becomes further distorted, or if physically impossible artifacts appear, then we may reject the hypothesis that we have correctly filled the lost or excised portion of Π_n. In general, of course, it may be virtually impossible to posit the lost portion. There is another approach, however.

Because the sense of the original image is retained, it may be possible to make reasonable and useful assumptions about portions of P_n which are visible within the remaining Π_n. One such common class is that of constant, or near constant, texture such as a clear sky, for example. Given that we can correctly assume what the some

[14] The reader is referred to Chapter 8, "Use of the 1–D Hadamard Transform in the Computation of a 2–D Hadamard Transform."

of the retained portions of Π_n should transform to, we may be able to fill in the missing information.

As an example, consider the series of pictures in Figure 30. Figure 30a is the original P_n. The image is 256x256 pixels with each pixel capable of 256 gray levels. Figure 30b is Π_n with $\alpha = 0.9$. Note that the original scene is still interpretable. Figure 30c is the excised Π_n (The bottom 57x256 values are set to zero. This is approximately 22% of the picture.) We note that the sky appears cloudless and therefore we have a very simple texture. We can attempt to reconstruct the missing pixels by solving for those values that will minimize the variance of the detransformed sky pixels. The classical and obvious way to do this would be to use the method of least squares. This would require us to solve for 14,592 unknowns in 32,768 equations corresponding to the sky pixels — a computationally challenging task.

Fortunately we were able to devise an iterative method with rapid convergence. The result of the algorithm applied to the example for 50 iterations is shown in Figure 30d.

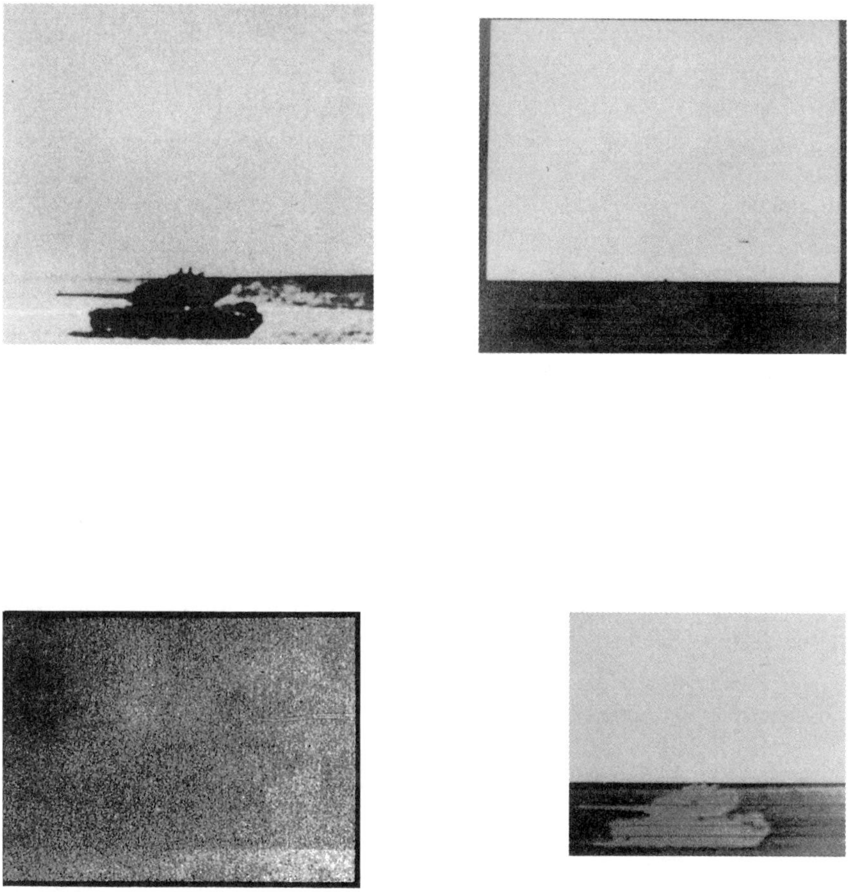

Figure 30 P_n — upper left (30a);
Π_n — upper right (30b);
Excised Π_n — lower left (30c);
Estimated P_n after 50 Iterations — lower right (30d).

The iterative algorithm bears close resemblance to work done on image reconstruction using the method of projection onto convex sets (POCS). [See, for example, Youla and Webb (1983); Sezan and Stark (1983).] The naturalness-preserving transform is a convex function and Euclidean-norm reducing. Restoration by POCS methods essentially trade *a priori* information for incomplete information. A solution, defined as a fixed point, ie, a vector left invariant under a suitable operation, is sought by successive iterations of the suitable operation. Convergence is guaranteed due to the convex nature of the range space. Our iterative technique is, however, quite different from the iterative techniques usually associated with POCS restorative methods.

Our technique is to iteratively compute (at the i-th stage)

$$
\begin{aligned}
P_n^{(i)} &= \Psi_n\left(\frac{\alpha}{2\alpha-1}\right)\widehat{\Pi}_n^{(i-1)}\Psi_n\left(\frac{\alpha}{2\alpha-1}\right) \\
\Pi_n^{(i)} &= \Psi_n(\alpha)\widehat{P}_n^{(i)}\Psi_n(\alpha)
\end{aligned}
\tag{301}
$$

where $\widehat{\Pi}_n^{(i)}$ and $\widehat{P}_n^{(i)}$ are obtained from $\Pi_n^{(i)}$ and $P_n^{(i)}$ by substituting the known values, ie, the non excised values in Π_n will be substituted at each stage in $\Pi_n^{(i)}$ to obtain $\widehat{\Pi}_n^{(i)}$; likewise the posited known values in P_n will be substituted into $P_n^{(i)}$ to obtain $\widehat{P}_n^{(i)}$.

The algorithm converges. To see this we first rewrite the iterative scheme in a more conventional form. Using (287) and (293) we have

$$
\Pi_n = CDC^T P_n CDC^T \tag{302}
$$

or

$$
D(C^T P_n C) = (C^T \Pi_n C) D^{-1} . \tag{303}
$$

As preparation, let

$$
\begin{aligned}
D &= diag(d_1, d_2, ..., d_N) \\
Q &= (q_{ij}) = C^T P_n C \\
R &= (r_{ij}) = C^T \Pi_n C
\end{aligned}
\tag{304}
$$

where the entries q_{ij} and r_{ij} are linear functions of entries in P_n and Π_n respectively. We now define the two vectors

$$
\begin{aligned}
q^T &= (q_{11}q_{12}\cdots q_{1N}q_{21}q_{22}\cdots q_{2N}\cdots q_{N1}\cdots q_{NN}) \\
r^T &= (r_{11}r_{12}\cdots r_{1N}r_{21}r_{22}\cdots r_{2N}\cdots r_{N1}\cdots r_{NN}) .
\end{aligned}
\tag{305}
$$

From (303) it is clear that

$$
D_1 q = D_2 r \tag{306}
$$

where

$$
D_1 = diag(d_1, \cdots, d_1, d_2, \cdots, d_2, \cdots, d_N, \cdots, d_N) \tag{307}
$$

and

$$
D_2 = diag\left(\frac{1}{d_1}, \frac{1}{d_2}, \cdots, \frac{1}{d_N}, \frac{1}{d_1}, \frac{1}{d_2}, \cdots, \frac{1}{d_N}, \frac{1}{d_1}, \frac{1}{d_2}, \cdots, \frac{1}{d_N}\right) . \tag{308}
$$

Note that q and r are vectors and are linear functions of entries in P_n and Π_n. Therefore we can write

$$D_1 X a = D_2 X b \tag{309}$$

where

$$\begin{aligned} a^T &= (\pi_{11}\pi_{12}\cdots\pi_{1N}\pi_{21}\cdots\pi_{2N}\cdots\pi_{N1}\cdots\pi_{NN}) \\ b^T &= (b_{11}b_{12}\cdots b_{1N}b_{21}\cdots b_{2N}\cdots b_{N1}\cdots b_{NN}) \end{aligned} \tag{310}$$

and the matrix X is

$$\begin{aligned} X &= C^T \otimes C^T \\ &= \begin{pmatrix} c_{11}C^T & c_{21}C^T & \cdots & c_{N1}C^T \\ c_{12}C^T & c_{22}C^T & \cdots & c_{N2}C^T \\ \vdots & \vdots & & \vdots \\ c_{1N}C^T & c_{2N}C^T & \cdots & c_{NN}C^T \end{pmatrix} \\ &= (x_{ij})_{N^2,N^2} . \end{aligned} \tag{311}$$

[See, for example, Gonzalez and Wintz (1977).]

Note that X is an orthogonal matrix because C is an orthogonal matrix. With these relationships, (310) can be written as

$$\begin{aligned} a &= X^T(D_3 X b) \\ b &= X^T(D_3^{-1})Xa \end{aligned} \tag{312}$$

where

$$D_3 = D_1^{-1} D_2 . \tag{313}$$

From the above it is evident that the iterative process may be implemented in terms of (312). Further, the iterative solutions obtained by using (301) or (312) will be identical. Let a_{U_1} and b_{U_2} be vectors consisting of U_1 and U_2 unknown entries from a and b respectively. Also, let a_{K_1} and b_{K_2} be the vectors consisting of K_1 and K_2 known entries from a and b respectively. From (312) we may write

$$\begin{aligned} a_{U_1} &= X_{U_1}^T D_3 X_{U_2} b_{U_2} + X_{U_1}^T D_3 X_{K_2} b_{K_2} \\ b_{U_2} &= X_{U_2}^T D_3^{-1} X_{U_1} a_{U_1} + X_{U_2}^T D_3^{-1} X_{K_1} a_{K_1} \end{aligned} \tag{314}$$

where $X_{U_i}^T$ and X_{K_i} are matrices obtained from X^T and X respectively, to determine a_{U_1} and b_{U_2} in terms of b_{U_2}, b_{K_2}, a_{U_1} and a_{K_1}. We may now write (314) in matrix form as

$$\begin{pmatrix} a_{U_1} \\ b_{U_2} \end{pmatrix} = \begin{pmatrix} 0 & X_{U_1}^T D_3 X_{U_2} \\ X_{U_2}^T D_3^{-1} X_{U_1} & 0 \end{pmatrix} \begin{pmatrix} a_{U_1} \\ b_{U_2} \end{pmatrix} + \begin{pmatrix} known \\ vector \end{pmatrix} \tag{315}$$

which is of the form

$$v = Gv + t \tag{316}$$

where t is a known vector, v is an unknown vector, and G is a known matrix.

It is clear that the iteration process can be identified by $v(n+1) = Gv(n) + t$ and the solution converges provided that the eigenvalues of G, $\{\lambda_i\}$, satisfy the constraint

$$|\lambda_i| < 1 \tag{317}$$

as specified in Varga (1962).

Let us now investigate the eigenvalues of G. The characteristic polynomial, $|\lambda I - G|$, is $|\lambda I_U - G| = \begin{pmatrix} \lambda I_{U_1} & -X_{U_1}^T D_3 X_{U_2} \\ -X_{U_2}^T D_3^{-1} X_{U_1} & \lambda I_{U_2} \end{pmatrix}$ where I_U, I_{U_1}, and I_{U_2} are identity matrices of order $U = U_1 + U_2$, U_1, and U_2 respectively. Without loss of generality, assume $U_2 \geq U_1$. Now

$$|\lambda I_U - G| = \lambda^{U_1 - U_2} \left| \lambda^2 I_{U_2} - X_{U_2}^T D_3^{-1} X_{U_1} X_{U_1}^T D_3 X_{U_2} \right| . \tag{318}$$

Noting that $D_3^{-1} X_{U_1} X_{U_1}^T D_3$ and $X_{U_1}^T D_3 D_3^{-1} X_{U_1} = I_{U_1}$ have the same eigenvalues, excepting that $D_3^{-1} X_{U_1} X_{U_1}^T D_3$ has an extra $N^2 - U_1$ zero eigenvalues [see Wilkinson (1965), p. 54], it follows [see Frame (1964)] that

$$\begin{aligned} D_3^{-1} X_{U_1} X_{U_1}^T D_3 &= L^T \begin{pmatrix} J_{U_1} & S \\ 0 & 0_{N^2 - U_1} \end{pmatrix} L \\ &= L^T J L \end{aligned} \tag{319}$$

where L is an orthogonal matrix, J and J_{U_1} are upper triangular matrices with unity diagonal entries in J_{U_1}, S a matrix not of interest here, and $0_{N^2 - U_1}$ is a null square matrix of order $N^2 - U_1$. Note that the order of the null matrix follows from the fact that the matrix $D_3^{-1} X_{U_1} X_{U_1}^T D_3$ has a rank of U_1 and it has U_1 unity eigenvalues. Using (319) we may write

$$X_{U_2}^T D_3^{-1} X_{U_1} X_{U_1}^T D_3 X_{U_2} = X_{U_2}^T L^T J L X_{U_2} . \tag{320}$$

We express

$$\begin{aligned} X_{U_2}^T L^T &= \begin{pmatrix} Y_{U_2}^T & Y_{N^2 - U_2}^T \end{pmatrix} \\ &= Y^T \end{aligned} \tag{321}$$

where Y_{U_2} is a square matrix of order U_2 and $Y_{N^2 - U_2}$ is a rectangular matrix of order $\left(N^2 - U_2\right) \times U_2$. Now (320) may be expressed as

$$X_{U_2}^T D_3^{-1} X_{U_1} X_{U_1}^T D_3 X_{U_2} = Y_{U_2}^T J Y_{U_2} . \tag{322}$$

Since Y in (321), and therefore Y_{U_2}, may be considered as submatrices of an orthogonal matrix, it follows that the spectral norm $||Y_{U_2}|| < 1$. The spectral norm (Varga, 1962) is subject to

$$\left\| Y_{U_2}^T J Y_{U_2} \right\| \leq \left\| Y_{U_2}^T \right\| \, ||J|| \, ||Y_{U_2}|| . \tag{323}$$

We note that the spectral radius of J is unity and so it follows that

$$\left\| Y_{U_2}^T J Y_{U_2} \right\| \leq 1 \tag{324}$$

which demonstrates that the magnitudes of all the eigenvalues of $X_{U_2}^T D_3^{-1} X_{U_1} X_{U_1}^T D_3 X_{U_2}$ are less than unity. This implies that all eigenvalues of G in (316) have magnitude less than unity which proves the convergence of the iterative process.

References

N. Ahmed and K. R. Rao (1975), Orthogonal Transforms for Digital Signal Processing, Springer-Verlag.

K. Beauchamp (1984), Applications of Walsh and Related Functions, Academic Press.

M. Bellanger (1984), Digital Processing of Signals, John Wiley & Sons.

J. B. Buehler and K. J. Miescke (1981), "On (n-1)-Wise and Joint Independence and Normality of n Random Variables: An Example," Communications Statistical Theoretical Methods, A10(9), pp. 927–930.

E. Carne (1965), Artificial Intelligence Techniques, Macmillan and Co.

C. K. Chow (1961), "On the Characterization of Threshold Functions," Proceedings of the IEEE Symposium on Switching Theory and Logic design, pp. 34–38.

M. Cohn and A. Lempel (1977), "On Fast M-Sequence Transforms," IEEE Transactions on Information Theory, January, pp. 135–137.

J. W. Cooley and J. W. Tukey (1965), "An Algorithm for Machine Computation of Complex F-Series," Mathematics of Computation, Vol. 19, pp. 297–301.

R. Craigen and W. Wallis (1993), "Hadamard Matrices: 1893–1993," Congressus Numerantium, Vol. 97, pp. 99–129.

M. L. Dertouzos (1965), Threshold Logic: A Synthesis Approach," USA Research Monograph 32, MIT Press.

M. F. Driscoll (1978), "On Pairwise and Mutual Independence: Characterization of Rectangular Distributions," Journal of the American Statistical Association, Vol. 73, pp. 432–3.

C. R. Edwards (1974), "Application of the Rademacher-Walsh Transform to Boolean Function Classification and Threshold Logic Synthesis," IEEE Transactions on Computers, Vol. C-24, pp. 48–62.

J. S. Frame (1964), "Matrix Functions and Applications," IEEE Spectrum, Vol. 1, April, pp. 102–108.

R. Gold (1966), "Characteristic Linear Sequences and Their Coset Functions," Journal SIAM, Vol. 14, No. 5, September, pp. 980–985.

S. W. Golomb (1959), "On the Classification of Boolean Functions," IRE Transactions on Circuit Theory, Vol. CT-6, pp. 176–186.

S. Golomb (1967), Shift Register Sequences, Holden-Day.

R. C. Gonzalez and P. Wintz (1977), Digital Image Processing, Addison-Wesley.

R. C. Gonzalez and P. Wintz (1987), Digital Image Processing, Second Edition, Addison-Wesley.

M. Harwit and N. J. A. Sloane (1979), Hadamard Transform Optics, Academic Press.

A. A. Hassan, J. E. Hershey, J. Schroeder, G. R. L. Sohie and R. K. R. Yarlagadda (1994), The Elements of System Design, Academic Press.

K. W. Henderson (1970), "Comment on 'Computations of the Fast Walsh-Fourier Transform'," IEEE Transactions on Computers, September, pp. 850–851.

J. E. Hershey (1979), "Comma-Free Synchronization of Binary Counters," IEEE Transactions on Information Theory, Vol. IT-25, No. 6, November pp. 724–725.

J. E. Hershey, G. J. Saulnier and N. Al-Dhahir (1996), "New Hadamard basis," Electronics Letters, Vol. 32, No. 5, pp. 429–430, 29 February.

J. E. Hershey (1980), "A Spectrally Preconditioned Threshold Logic Unit," Proceedings of the 5th International Conference on Pattern recognition, pp. 680–684.

J. E. Hershey and W. F. Lawrence (1981), "Counter Synchronization of Binary Using the Thue-Morse Sequence and PSK," IEEE Transactions on Communications, Vol. COM-29, No. 1, January pp. 79–80.

J. Hershey and R. Yarlagadda (1983), "A New Class of Image Transforms," Proceedings of the 1983 IEEE Military Communications Conference.

J. E. Hershey and R. K. R. Yarlagadda (1986), Data Transportation and Protection, Plenum Press.

J. Hershey and J. Tiemann (1996), "Random Parity Coding," International Conference on Communications 1996 (ICC'96), Vol. 1, pp. 122–126.

A. S. Householder and K. Fox, "Determination of Eigenvectors of Symmetric Idempotent Matrices," Journal of Computational Physics, Vol. 8, pp. 292–294.

S. Hu (1965), Threshold Logic, University of California Press.

S. L. Hurst (1973), "Application of Chow Parameters and Rademacher-Walsh Matrices in the Synthesis of Binary Functions," Computer Journal, Vol. 16, pp. 167–173.

M. G. Karpovsky (1976), Finite Orthogonal Series in the Design of Digital Devices, John Wiley and Sons.

H. Kremer (1973), "Eigen Analysis of Walsh Matrices," in *Proceedings of the Symposia Held at Hatfield Polytechnic*, Hatfield, Hertfordshire, England, pp. 1a-6a.

M. H. Lee and M. Kaveh (1986), "Fast Hadamard Transform Based on a Simple Matrix factorization," IEEE Transactions on Acoustics, Speech, and Signal Processing, Vol. ASSP-34, No. 6, December, pp. 1666–1667.

S. -Y. Kung (1984) "On Supercomputing with Systolic/Wavefront Array Processors," Proceedings of the IEEE, Vol. 72, Number 7, pp. 867–884.

M. H. Lee and Y. Yasuda (1990), "Simple Systolic Array Algorithm for Hadamard Transform," Electronics Letters, 30th August, Vol. 26, No. 18, pp. 1478–1480.

A. Lempel (1975), "Matrix Factorization Over GF(2) and Trace-Orthogonal Bases of GF(2^n)," SIAM Journal of Computation, Vol. 4, No. 2, June, pp. 175–186.

A. Lempel (1979), "Hadamard and M-Sequence transforms are permutationally similar," Applied Optics, Vol. 18, No. 24, 15 December, pp. 4064–4065.

S. G. Mallat (1989), "A Theory for Multiresolution Signal Decomposition: The Wavelet Representation," IEEE Transactions on Pattern Analysis and Machine Intelligence, 11, 674–693.

H. B. Mann (1968), editor, Error Correcting Codes, John Wiley & Sons, Inc.

A. W. Naylor and G. R. Sell (1982), Linear Operator Theory in Engineering and Science, Springer-Verlag.

N. Nilsson (1965), Learning Machines, McGraw-Hill.

J. D. Olsen, R. A. Scholtz and L. R. Welch (1982), "Bent-function Sequences," IEEE Transactions on Information Theory, Vol. IT-28, pp. 858–864.

A. V. Oppenheim and R. W. Schafer (1989), Discrete-Time Signal Processing, Prentice-Hall, Englewood Cliffs, New Jersey.

W. Press et al (1992) Numerical Recipes in Fortran: The Art of Scientific Computing 2nd. edition, Cambridge.

J. Pearl (1971), "Application of Walsh Transform to Statistical Analysis," IEEE Transactions on Systems, Man, and Cybernetics, Vol. SMC-1, No. 2, April, pp. 111–119.

W. W. Peterson and E. J. Weldon, Jr. (1972), Error-Correcting Codes, second edition, The MIT Press.

P. D. Picton (1981), "Realisation of multithreshold logic networks using the Rademacher-Walsh transform," Proceedings of the IEE, Vol. 128, pp. 107–113.

P. A. Regalia and S. K. Mitra (1989), "Kronecker Products, Unitary Matrices and Signal Processing Applications," SIAM Review, Vol. 31, No. 4, December, pp. 586–613.

H. L. Resnikoff (1990), "Weierstrass Functions and Compactly Supported Wavelets," Aware Technical Report No. AD900810.

F. Rosenblatt (1962), Principles of Neurodynamics, Spartan Books.

O. Rothaus (1976), "On 'Bent' Functions," Journal of Combinatorial Theory, Series A, Vol. 20, No. 3, May, pp. 300–305.

M. I. Sezan and H. Stark (1983), "Image Restoration by the Method of Convex Projections: Part 2 — Applications and Numerical results," IEEE Transactions on Medical Imaging, Vol. MI-1, pp. 95-101.

J. M. Stoyanov (1987), Counterexamples in Probability, John Wiley & Sons.

G. Strang (1989), "Wavelets and Dilation Equations: A Brief Introduction," SIAM Review, Vol. 31, No. 4, pp. 614–627.

W. Stahnke (1973), "Primitive Binary Polynomials," Mathematics of Computation, Vol. 27, No. 124, October, pp. 977–980.

J. Sylvester (1867), "Thoughts on Inverse Orthogonal Matrices, simultaneous Sign-successions, and Tessellated Pavements in two or more colours, with applications to Newton's Rule, Ornamental Tile-work, and the Theory of Numbers," Phil. Mag. Series 4, No. 34, pp. 461–475.

R. Titsworth (1964), "Optimal Ranging Codes," IEEE Transactions on Space Electronics and Telemetry, Vol. SET-10, March, pp. 19–30.

R. S. Varga (1962), Matrix Iterative Analysis, Prentice-Hall.

M. Vetterli (1983), "Tree Structures for Orthogonal Transforms and Application to the Hadamard Transform," Signal Processing, Vol. 5, pp. 473–484.

G. Weathers (1972), "Statistical Properties of Filtered Pseudo-Random Digital Sequences, Sperry Rand Corporation, Technical Report No. SP-275-0599, January. (Prepared for NASA George C.

Marshall Space Flight Center, Huntsville, AL.)

J. H. Wilkinson (1965), The Algebraic Eigenvalue Problem, Clarendon.

S. G. Wilson and M. Lakshman (1968), "Autocorrelation and power spectrum of Hadamard signalling," IEE Proceedings, Vol. 135, Pt. F, No. 3, June 1968, pp. 258–261.

R. Yarlagadda and J. Hershey (1981), "Architecture of the fast Walsh-Hadamard and Fast Fourier transforms with charge transfer devices," International Journal of Electronics, Vol. 51, No. 5, pp. 669–681.

R. Yarlagadda and J. Hershey (1982a), "A Note on the Eigenvectors of Hadamard Matrices of Order 2^n," Linear Algebra and Its Applications, Vol. 45, pp. 43–53.

R. Yarlagadda and J. E. Hershey (1982b), "Benchmark Synchronization of m-Sequences," Electronics Letters, 21 January, Vol. 18, No. 2, pp. 68–69.

R. Yarlagadda and J. Hershey (1984), "Spectral Properties of the Thue-Morse Sequence," IEEE Transactions on Communications, COM-32, 974–977.

R. Yarlagadda and J. Hershey (1985), "A Naturalness-Preserving Transform for Image Coding and Reconstruction," IEEE Transactions on Acoustics, Speech, and Signal Processing, Vol. ASSP-33, No. 4, August, pp. 1005–1012.

R. Yarlagadda and J. E. Hershey (1989), "Analysis and Synthesis of Bent Sequences," IEE Proceedings, Vol. 136, Pt. E, No. 2, March, pp. 112–123.

R. Yarlagadda and J. E. Hershey (1990), "Autocorrelation Properties of the Thue-Morse Sequence and Their Use in Synchronization," IEEE Transactions on Communications, COM-38, No. 12, December pp. 2099–2102.

F. Yates (1937), The Design and Analysis of Factorial Experiments, Imperial Bureau of Soil Science, Harpenden, England.

D. C. Youla and H. Webb (1983), "Image Restoration by the Method of Convex Projections: Part 1 — Theory," IEEE Transactions on Medical Imaging, Vol. MI-1, pp. 81–94.

N. Zierler (1959), "Linear Recurring Sequences," Journal SIAM, Vol. 7, No. 1, March, pp. 31–48.

Index